# Lecture Notes in Mathematics

Edited by A. Dold and B. Eckmann

1383

D.V. Chudnovsky G.V. Chudnovsky
H. Cohn M.B. Nathanson (Eds.)

# Number Theory

A Seminar held at the Graduate School and
University Center of the City University
of New York 1985–88

Springer-Verlag
Berlin Heidelberg New York London Paris Tokyo Hong Kong

**Editors**

David V. Chudnovsky
Gregory V. Chudnovsky
Department of Mathematics, Columbia University
New York, NY 10027, USA

Harvey Cohn
Department of Mathematics, (CUNY), City College
New York, NY 10031, USA

Melvyn B. Nathanson
Provost and Vice President for Academic Affairs
Lehman College (CUNY)
Bronx, NY 10468, USA

Mathematics Subject Classification (1980): 10-06

ISBN 3-540-51549-6 Springer-Verlag Berlin Heidelberg New York
ISBN 0-387-51549-6 Springer-Verlag New York Berlin Heidelberg



Printing and binding: Druckhaus Beltz, Hemsbach/Bergstr.
2146/3140-543210 – Printed on acid-free paper

# INTRODUCTION

This is the fourth volume of papers presented at the New York Number Theory Seminar. Since 1982 the Seminar has been meeting every Tuesday afternoon during the academic year at the Graduate School and University Center of the City University of New York. The goal of the Seminar is to provide a forum for the exposure of new results in number theory and allied fields in the New York metropolitan area. Mathematicians who plan to be in New York and would like to attend or lecture in the Seminar are encouraged to contact the organizers.

## TABLE OF CONTENTS

D. BUMP, J. HOFFSTEIN, Some Conjectured Relationships Between Theta Functions and Eisenstein Series on the Metaplectic Group.......................................... 1

D.V. CHUDNOVSKY, G.V. CHUDNOVSKY, Computational Problems in Arithmetic of Linear Differential Equations. Some Diophantine Applications.......................................... 12

H. COHN, Iteration of Two-Valued Modular Equations..... 50

D. GOSS, Report on Transcendency in the Theory of Function Fields.......................................... 59

D. HAJELA, Exponential Sums and Faster Than Nyquist Signaling.......................................... 64

A. HILDEBRAND, Some New Applications of the Large Sieve.......................................... 76

W. HOYT, Elliptic Fibering of Kummer Surfaces.......... 89

M.I. KNOPP, Recent Developments in the Theory of Rational Period Functions.......................................... 111

M.B. NATHANSON, Additive Problems in Combinatorial Number Theory.......................................... 123

R. RILEY, Growth of Order of Homology of Cyclic Branched Covers of Knots.......................................... 140

A. SÁRKÖZY, Hybrid Problems in Number Theory........... 146

A.H. STEIN, Binomial Coefficients Not Divisible by a Prime.......................................... 170

M.E. SWEEDLER, Positive Characteristic Calculus and Icebergs.......................................... 178

N. SUWA, N. YUI, Arithmetic of Certain Algebraic Surfaces over Finite Fields.......................................... 186

# SOME CONJECTURED RELATIONSHIPS BETWEEN THETA FUNCTIONS AND EISENSTEIN SERIES ON THE METAPLECTIC GROUP

BY DANIEL BUMP AND JEFFREY HOFFSTEIN

**This research was supported by NSF Grants # DMS 8612896 and # DMS 8519916. We would like to express our thanks to S. J. Patterson for many helpful discussions.**

Suzuki [6] considered the Fourier coefficients of a theta function on the four-fold cover of $GL(2)$. Despite courageous efforts, he was only able to obtain partial information about these coefficients. This failure was explained by the work of Kazhdan and Patterson [2], who showed that if $r < n-1$, the methods of Hecke theory only yield partial information about the Fourier coefficients of theta functions on the $n$-fold cover of $GL(r)$, owing to the fact that the local representations do not have unique Whittaker models. Nevertheless, in the special case $r = 2$, $n = 4$, Patterson [4], [5] was able to formulate a conjecture which would specify the unknown coefficients up to sign, as square roots of Gauss sums. This work inspired a further paper of Suzuki [7], but unfortunately it is not clear to us precisely what is proved in this latter paper.

We shall review the evidence of Patterson, and develop further evidence of our own. Essentially, our theme is that if Patterson's conjecture is true, there are identities between various Dirichlet series which arise as Fourier coefficients of Eisenstein series or as Rankin-Selberg convolutions. By comparing the functional equations and the locations of the poles of these Dirichlet series, we become convinced that the conjecture must be true. Indeed, in some sense the conjecture must be true "on average" owing to the locations of these poles.

We shall finally state a very general conjecture asserting the equality of two Rankin-Selberg convolutions of theta functions. These convolutions may also be interpreted (conjecturally) as a Fourier coefficient of an Eisenstein series on the metaplectic group, generalizing a key relation in our discussion of Patterson's conjecture. This conjecture allows us to predict many values of (and relations between values of) Fourier coefficients of theta functions beyond what is predicted by the theory of Kazhdan and Patterson. We also conjecture that the Rankin-Selberg convolution of a metaplectic cusp form with a theta function may be interpreted as the Fourier coefficient of an Eisenstein series.

Because the forms which we shall consider are automorphic with respect to congruence subgroups, which have multiple cusps, the various Dirichlet series which we shall consider will need congruence conditions. These congruence conditions also complicate the functional equations of these Dirichlet series. As one sees for example in [6], keeping track of these nuances involves some rather tedious bookkeeping. We shall not state these congruence conditions explicitly, because they are a distraction, and because we have not worked out all the details arising from them. Nor shall we state the functional equations precisely, or compute the Gamma factors which go with them. *Thus many of the formulas contained herein should be taken as suggestive rather than strictly truthful.* We hope that this transgression will be excused on the grounds that it allows us to tell our story more freely.

Let $F$ be a number field which, for simplicity, we assume to be totally complex. Let $n$ be a fixed positive integer, and assume that the group $\mu_n$ of $n$-th roots of unity in $\mathbf{C}$ is contained in $F$. Let $\left(\frac{a}{b}\right)$ be the $n$-th power residue symbol for the field $F$, which is defined for coprime $a$, $b$ in the ring $O$ of integers in $F$, and takes values in $\mu_n$. Among the properties of this symbol which we shall need, it satisfies

$$\left(\frac{aa'}{b}\right) = \left(\frac{a}{b}\right)\left(\frac{a'}{b}\right), \qquad \left(\frac{a}{bb'}\right) = \left(\frac{a}{b}\right)\left(\frac{a}{b'}\right),$$

$$\left(\frac{a}{b}\right) = \left(\frac{a'}{b}\right) \qquad \text{if } a \equiv a' \bmod b,$$

and the reciprocity law

$$\left(\frac{a}{b}\right) = (b,a)\left(\frac{b}{a}\right),$$

where $(b,a)$ is a "Hilbert symbol".

We shall eventually be concerned with the particular case where $F = Q(i)$. In this case, any ideal which is prime to $\lambda = 1 + i$ has a unique generator which is congruent to 1 mod $\lambda^3$. We shall always use that generator. Thus if $p$ is a prime, we always assume that $p \equiv 1 \bmod \lambda^3$. In this case (for such $a$ and $b$), the Hilbert symbol

$$(a,b) = (-1)^{\frac{1}{2}(Na-1)\,\frac{1}{2}(Nb-1)}.$$

Kubota proved that

$$\kappa\begin{pmatrix} a & b \\ c & d \end{pmatrix} = \left(\frac{c}{d}\right)$$

defines a character of the congruence subgroup $\Gamma(n^2)$ of $SL(2,O)$. Furthermore, if we embed $SL(2,O)$ into $SL(r,O)$ by sending a $2 \times 2$ matrix into the upper right hand corner of an $r \times r$ matrix (with ones elsewhere on the diagonal), then the Kubota character $\kappa$ extends to a character $\kappa$ of a congruence subgroup of $SL(r,O)$. We shall be concerned with *metaplectic forms*, which are automorphic forms on such a congruence subgroup, formed with the Kubota character.

Firstly, let us consider automorphic forms on $SL(2)$ which satisfy

$$\phi(\tau) = \kappa(\gamma)\,\phi(\gamma\tau)$$

for $\gamma$ in the congruence subgroup $\Gamma(n^2)$ and $\tau$ in the homogeneous space, which is product of hyperbolic 3-spaces. As was noted by Hecke, Wohlfahrt and Shimura, the theory of Hecke operators for such forms is different from the theory of Hecke operators for nonmetaplectic forms. Let us review such a theory (a more careful treatment may be found in Bump and Hoffstein [1]). We assume that $\phi$ has a Fourier expansion

$$\phi(\tau) = \sum_m a(m)\, Nm^{-1/2}\, W(m\tau),$$

where $W$ is a "Whittaker function" (essentially a product of $K$-Bessel functions).

The Hecke operators are double cosets whose elementary divisors are $n$-th powers. If $p$ is a prime of $\mathcal{O}$ which does not divide $n$, decompose the double coset

$$\Gamma\xi\Gamma = \bigcup_i \xi_i, \qquad \xi_i = \gamma_i \xi \delta_i,$$

where

$$\xi = \begin{pmatrix} 1 & \\ & p^n \end{pmatrix}, \qquad \gamma_i, \delta_i \in \Gamma.$$

Then define

$$(T_{p^n}\phi)(\tau) = \sum_i \kappa(\gamma_i)\, \kappa(\delta_i)\, \phi(\xi_i \tau).$$

As in Bump and Hoffstein [1], one may explicitly compute the decomposition of the double coset, and consequently, the effect of the Hecke operators on the Fourier coefficients. In the case $n = 4$, the $m$-th Fourier coefficient of $T_{p^4}\phi$ is (to oversimplify somewhat)

$$A(m) = Np^2\, a\left(\frac{m}{p^4}\right) + Np\,(m,p)\, g_3\left(\frac{m}{p^2}, p\right)\, a\,(m/p^2))$$

$$+ Np\,(m,p^2)\, g_2\left(\frac{m}{p}, p\right)\, a(m) + Np\,(m,p^3)\, g_1(m,p)\, a(mp^2) + Np^2\, a(mp^4).$$

Here the Gauss sums are defined by

$$g_r(m,p) = \sum_{k \bmod p} \left(\frac{k}{p}\right)^r e\left(\frac{mk}{p}\right),$$

in terms of the fourth power residue symbol, where $e$ is a certain fixed character of $F$ mod $\mathcal{O}$. We interpret $a(m) = 0$ and $g_r(m,p) = 0$ if $m$ is not integral.

The *theta function* $\theta$ is a residue of a certain Eisenstein series, which is automorphic with respect to the fourth power Kubota symbol. It is an eigenfunction of the Hecke operators with $T_{p^4}\theta = Np^2\, \lambda_p\, \theta$, where $\lambda_p = Np^{1/2} + Np^{-1/2}$, and so its Fourier coefficients $\tau(m)$ satisfy

(1) $(Np^{1/2}+Np^{-1/2})\tau(m)$

$$= \tau\left(\frac{m}{p^4}\right) + Np^{-1}\ (m,p)\, g_3\left(\frac{m}{p^2},p\right)\,\tau\left(\frac{m}{p^2}\right)$$

$$+ Np^{-1}\ (m,p^2)\, g_2\left(\frac{m}{p},p\right)\ \tau(m) + Np^{-1}\ (m,p^3)\, g_1(m,p)\,\tau(mp^2) + \tau(mp^4).$$

Furthermore, we have the following *Periodicity Theorem*, a simple but rather deeper fact, which was proved in complete generality by Kazhdan and Patterson [**2**]. This is the fact that $\tau(m)$ depends only on $m$ modulo fourth powers:

(2) $$\tau(h^4 m) = Nh^{1/2}\,\tau(m).$$

These relations tell us quite a bit about $\tau(m)$. If the theta function is normalized so that $\tau(1)=1$, with $m=1$, (1) and (2) imply that

$$Np^{-1/2} + Np^{1/2} = Np^{-1}\, g_1(1,p)\tau(p^2) + Np^{1/2},$$

and so

$$\tau(p^2) = Np^{-1/2}\overline{g_1(1,p)}.$$

Similarly, taking $m=p^3$ in (1) gives

$$\tau(p^3) = 0.$$

On the other hand, taking $m=p$ in (1) reduces to a tautology, since the quadratic Gauss sum $g_2(1,p) = Np^{1/2}$. The relations (1) and (2) do not imply anything about the values of $a(p)$. Still, we may sometimes show that $a(m)=0$ for squarefree $m$ (the coefficients are *not* expected to be multiplicative!) For example, taking $m=pq$, where $q$ is a different prime from $p$, we have

$$(Np^{1/2}+Np^{-1/2})\tau(pq) = Np^{-1}\, g_2(q,p)\,\tau(pq) + Np^{1/2}\,\tau(pq).$$

Since the quadratic Gauss sum

$$g_2(q,p) = \left(\frac{q}{p}\right)^2 Np^{1/2},$$

this implies that $\tau(pq) = 0$ if $p$ is a quadratic nonresidue modulo $q$, and more generally, it may be shown that if $m$ is squarefree, then

$$\tau(m) = 0 \qquad \text{if} \qquad \left(\frac{m_1}{m_2}\right)^2 = -1$$

for any factorization $m = m_1 m_2$.

These relations are essentially those found by Suzuki. To go beyond this, Patterson considered the Rankin-Selberg convolution of $\theta$ with itself. This is an integral of $\theta^2$ against a (quadratic metaplectic) Eisenstein series. This integral represents the Dirichlet series

$$\varsigma(4s-1) \sum \tau(m)^2 \, Nm^{-s},$$

where $\varsigma$ is the Dedekind zeta function of the field. This Dirichlet series has a functional equation under $s \mapsto 1-s$, and a pole at $s = \frac{3}{4}$. On the other hand, the Dirichlet series (first considered by Kubota)

$$\psi(s) = \sum g_1(1,m) \, Nm^{-s}$$

occurs in the Fourier coefficients of the quartic metaplectic Eisenstein series on $GL(2)$—the precise coefficient, which has a functional equation under $s \mapsto 1-s$, is $\varsigma(8s-3)\,\psi(2s)$, with a pole at $s = \frac{5}{8}$. Consequently, $\varsigma(4s-1)\,\psi(s+\frac{1}{2})$ also has a functional equation under $s \mapsto 1-s$, and a pole at $s = \frac{3}{4}$. Now Patterson made the remarkable observation that the assumption that

$$\sum \tau(m)^2 \, Nm^{-s} = \varsigma(4s-1)\,\overline{\psi\left(s+\tfrac{1}{2}\right)^2} \tag{3}$$

is consistent with everything which is known about $\tau(m)$. For example, after multiplying both sides by $\varsigma(4s-1)$, both sides have the same pole and functional equation (Patterson checked that the Gamma factors are the same). Moreover, the properties of $\tau$ which were found by Suzuki are consistent with this conjecture: the factor $\varsigma(4s-1)$ causes the coefficients on the right to be periodic, as predicted by the Periodicity Theorem (2), and, for example, if $m$ is square-free and admits a factorization $m = m_1 m_2$ with

$$\left(\frac{m_1}{m_2}\right)^2 = -1,$$

then cancellations cause the coefficient of $Nm^{-s}$ on the right to vanish. On the other hand, the other squarefree coefficients will not vanish—if the squarefree $m$ admits no such factorization into $m_1 m_2$, the conjecture implies that

$$\tau(m)^2 = 2^k \, Nm^{-1/2}\,\overline{g_1(1,m)}.$$

Here $k$ is the number of prime factors of $m$. Thus Patterson's conjecture determines all the Fourier coeffients of $\theta$, at least *up to sign*.

Furthermore, Patterson considered the convolution of $\theta$ with its *complex conjugate*. This is the integral of $|\theta|^2$ against a nonmetaplectic Eisenstein series. It represents the Dirichlet series

$$\zeta(2s) \sum |\tau(m)|^2 \, Nm^{-s}, \tag{4}$$

which has analytic continuation and a functional equation with respect to $s \mapsto 1-s$. There is a pole at $s = 1$ (there are also poles at $s = 0$, $\frac{1}{4}$ and $\frac{3}{4}$). The location of the pole is consistent with the *magnitude* of $\tau(m)^2$ predicted by the conjecture—for squarefree $m$, the conjecture predicts that $|\tau(m)|$ would be $2^k$ with probability $2^{1-k}$ and otherwise zero, where $k$ is the number of prime factors of $m$. We shall see later that if the conjecture is true, (4) is equal to a Dirichlet series which comes up in another context, and which does in fact have a functional equation and a simple pole at $s = 1$.

To go beyond this evidence of Patterson, let us consider an Eisenstein series on $GL(4)$. Specifically, let us define a function $I(\tau, s)$, where $\tau$ lies in $GL(4, \mathbf{C})/ZU(4)$ ($Z$ being the center of $GL(4, \mathbf{C})$), and $s$ is a complex parameter. Namely, any element of this homogeneous space has a representative of the form

$$\tau = \begin{pmatrix} y_1 y_2 y_3 & y_2 y_3 x_1 & y_3 x_4 & x_6 \\ & y_2 y_3 & y_3 x_2 & x_5 \\ & & y_3 & x_3 \\ & & & 1 \end{pmatrix}.$$

Then we let

$$I(\tau, s) = \theta \begin{pmatrix} y_1 & x_1 \\ & 1 \end{pmatrix} \theta \begin{pmatrix} y_2 & x_2 \\ & 1 \end{pmatrix} |y_1 y_2^2 y_3|^{2s},$$

where the $y_i$ are positive real numbers. Let $\Gamma_0(4)$ be the subgroup of matrices in $\Gamma(4)$ such that $2 \times 2$ block in the lower left hand corner consists of zeros. Then we have the following Eisenstein series:

$$E^*(\tau, s) = \zeta(8s - 7)\, \zeta(8s - 6)\, E(\tau, s),$$

$$E(\tau, s) = \sum_{\Gamma_0(4) \backslash \Gamma(4)} \kappa(\gamma)\, I(\gamma\tau, s).$$

This Eisenstein series has a functional equation with respect to $s \mapsto 2 - s$, with poles at $s = \frac{5}{4}, \frac{9}{8}, 1, \frac{7}{8}$, and $\frac{3}{4}$. Let us consider the Fourier coefficients. Specifically, let

$$w_1 = \begin{pmatrix} & & & 1 \\ & & 1 & \\ & \cdot^{\cdot^{\cdot}} & & \\ 1 & & & \end{pmatrix}.$$

The Fourier coefficients $D(s;n_1,n_2,n_3)$ are defined by

$$\int_{C/O}\cdots\int_{C/O} E\left(w_1\begin{pmatrix}1 & x_1 & x_4 & x_6\\ & 1 & x_2 & x_5\\ & & 1 & x_3\\ & & & 1\end{pmatrix}\tau\right) e(-n_1x_1-n_2x_2-n_3x_3)\,dx_1\cdots dx_6$$

$$= D(s;n_1,n_2,n_3)\,Nn_1^{-3/2}\,Nn_2^{-2}\,Nn_3^{-3/2}\,W\left(\begin{pmatrix}n_1n_2n_3 & & & \\ & n_2n_3 & & \\ & & n_3 & \\ & & & 1\end{pmatrix}\tau\right),$$

Where $D(s;n_1,n_2,n_3)$ is a certain Dirichlet series involving the coefficients $\tau(m)$. Full details of the determination of these Dirichlet series will be given elsewhere, but here we recapitulate the basic idea. (Indeed, it is necessary to do this simply in order to state the definition of $D(s,n_1,n_2,n_3)$.) A coset in $\Gamma_O\backslash\Gamma$ is given by the following data: If $\gamma$ is a matrix with the $i,j$-th entry being equal to $c_{ij}$, let $A_{ij}$, for $1\le i<j\le 4$ be the minor $c_{3i}c_{4j}-c_{3j}c_{4i}$. Then the coset of $\gamma$ in $\Gamma_O\backslash\Gamma$ is associated with the six numbers $A_{12}$, $A_{13}$, $A_{14}$, $A_{23}$, $A_{24}$ and $A_{34}$, which are coprime, and which satisfy

$$A_{12}A_{34}-A_{13}A_{24}+A_{14}A_{23}=0. \tag{5}$$

Conversely, given six coprime integers subject to the condition (5), there exists a coset having those numbers as minors. In computing the Fourier coefficients of the Eisenstein series, it is important to chose the coset representatives in a particular way. Specifically, let the $A_{ij}$ be given. The coefficients $c_{ij}$ are to be reconstructed as follow: let $A_4$ be the greatest common divisor of $A_{34}$, $A_{24}$ and $A_{14}$. Find $r$, $s$ and $t$ so that $A_4=rA_{34}+sA_{24}+tA_{14}$, and let $A_3=sA_{23}+tA_{13}$, $A_2=-rA_{23}+tA_{12}$, $A_1=-rA_{13}-sA_{12}$. Also, let $A_{234}$ be the greatest common divisor of $A_{34}$, $A_{24}$ and $A_{23}$. Find $R$, $S$ and $T$ so that $A_{234}=RA_{34}+SA_{24}+TA_{23}$, and let $A_{134}=SA_{14}+TA_{13}$, $A_{124}=-RA_{14}+TA_{12}$ and $A_{123}=-RA_{13}-SA_{12}$. It may be shown that $A_1$, $A_2$, $A_3$ and $A_4$ are coprime, and that $A_{123}$, $A_{124}$, $A_{134}$ and $A_{234}$ are coprime. Furthermore, we may choose the coset representative $\gamma$ so as to have bottom row $(A_1,A_2,A_3,A_4)$, and so that the bottom row of the involute ${}^{\iota}\gamma=w_1\,{}^{t}\gamma^{-1}\,w_1$ has bottom row $(-A_{123},A_{124},-A_{134},A_{234})$. This done, we may now describe $D(s;n_1,n_2,n_3)$: In fact, this is the Dirichlet series

$$\sum_{A_{34}}\ \sum_{\substack{A_{24},A_{14},A_{23},A_{13}\bmod A_{34}\\ A_{34}\mid A_{13}A_{24}-A_{14}A_{24}\\ A_{12},A_{13},A_{14},A_{23},A_{24},A_{34}\text{ coprime}\\ A_{234}^2\mid n_1A_{34},\,A_4^2\mid n_3A_{34}}} \left\{\kappa(\gamma)N(A_4A_{234}A_{34}^{-1})\,\tau\!\left(\frac{n_1A_{34}}{A_{234}^2}\right)\tau\!\left(\frac{n_3A_{34}}{A_4^2}\right)\right.$$
$$\left. e\!\left(n_1\frac{A_{134}}{A_{234}}+n_2\frac{A_{24}}{A_{34}}+n_3\frac{A_3}{A_4}\right)\right\}NA_{34}^{-2s}. \tag{6}$$

For the moment, we are only concerned with the coefficient where $n_1 = n_2 = n_3 = 1$. We see that

$$\zeta(8s-7)\,\zeta(8s-6)\,D(s;1,1,1)$$

has a functional equation with respect to $s \mapsto 2-s$, with a simple pole at $s = \frac{5}{4}$.

We have shown how Patterson deduced from the location of the pole of the Dirichlet series (4) that, on the average, the $\tau(m)$ have the same *magnitude* as predicted by his conjecture. Now let us show that the location of the pole of (6), with $n_1 = n_2 = n_3 = 1$, shows that on the average, the $\tau(m)$ have the right *arguments*. Thus, we seek to show that $\tau(p)^2 g_1(1,p)$ is, on the average, about $Np^{1/2}$. Indeed, we may calculate the coefficients in (6) more explicitly, and interestingly enough, like the series (4) and the left side of (3), they only involve the *squares* of the $\tau(p)$. Let us restrict ourselves to describing the $p$-part of the series (6) (with $n_1 = n_2 = n_3 = 1$), in other words, the sum of the coefficients of $Np^{-ks}$. Of course, one must compute all the coefficients, which we have done, but for the moment considering just the $p$-part will be sufficient to show what is happening. It is convenient to make the following change of variables: let $w = 2s - \frac{3}{2}$. Then the $p$-part is

$$1 + g_1(1,p)\,\tau(p)^2\,Np^{-\frac{1}{2}-w} + 2g_1(1,p)\,\tau(p^2)\,Np^{-\frac{1}{2}-2w} + g_1(1,p)\,\tau(p)^2\,Np^{-\frac{1}{2}-3w} + Np^{-4w}.$$

This is to have a functional equation with respect to $w \mapsto 1-w$, and a simple pole at $w = 1$. Since $|\tau(p)|$ is, on the average, constant (from the location of the pole of (4)), if the argument of $\tau(p)^2$ was not approximately the same as $\overline{g_1(1,p)}$, the pole of (6) would be to the *left* of $w = 1$. Thus the location of the pole shows that on the average, the argument of the $\tau(m)^2$ is consistent with the conjecture. Actually if $m$ is squarefree, and one assumes the conjecture, then the coefficient of $Nm^{-w}$ would be $2^k$ with probability $2^{1-k}$, where $k$ is the number of prime factors of $m$, and zero otherwise.

Now let us show that, if the conjecture is true, then the Dirichlet series (4) and (6) may actually be identified with known Dirichlet series having the correct functional equations and poles. Firstly, assuming the conjecture, the following identity may be established:

(7) $$\zeta(4w-1)\,\zeta(4w)\,D\big(\tfrac{1}{2}(w+\tfrac{3}{2});1,1,1\big) = \zeta(2w)\sum |\tau(m)|^2\,Nm^{-w}.$$

It follows from the general theory of Eisenstein series that the left hand side has simple poles at $s = 0$ and 1, and at $s = \frac{1}{4}$, and $\frac{3}{4}$. (The Eisenstein series itself also has a pole at $s = \frac{1}{2}$, but only the degenerate Fourier coefficients have poles—the left hand side of (7) has no pole at $s = \frac{1}{2}$.) The right hand side has the same poles.

We shall further show that assuming Patterson's conjecture, the above two Dirichlet series may be realized as the Fourier coefficient of an Eisenstein series on the two-fold cover of $GL(3)$. Specifically, if

$$\tau = \begin{pmatrix} y_1y_2 & y_2x_1 & x_3 \\ & y_2 & x_2 \\ & & 1 \end{pmatrix}, \qquad y_i > 0,$$

let us define

$$I_{\nu_1,\nu_2}(\tau) = y_1^{4\nu_1+2\nu_2}\, y_2^{2\nu_1+4\nu_2},$$

$$E(\tau,\nu_1,\nu_2) = \varsigma(6\nu_1-1)\,\varsigma(6\nu_2-1)\,\varsigma(6\nu_1+6\nu_2-3)\sum_{\Gamma_\infty(4)\backslash\Gamma(4)} \kappa(\gamma)^2\, I_{\nu_1,\nu_2}(\gamma\tau).$$

The Kubota symbol is squared to indicate that this Dirichlet series is made with *quadratic* symbols. The Eisenstein series $E(\tau,\nu_1,\nu_2)$ has functional equations with respect to

$$(\nu_1,\nu_2)\mapsto(\tfrac{2}{3}-\nu_2,\tfrac{2}{3}-\nu_1),$$

$$(\nu_1,\nu_2)\mapsto(\nu_1+\nu_2-\tfrac{1}{3},\tfrac{2}{3}-\nu_2),$$

$$(\nu_1,\nu_2)\mapsto(\tfrac{2}{3}-\nu_1,\nu_1+\nu_2-\tfrac{1}{3}),$$

$$(\nu_1,\nu_2)\mapsto(1-\nu_1-\nu_2,\nu_1),$$

$$(\nu_1,\nu_2)\mapsto(\nu_2,1-\nu_1-\nu_2).$$

The leading $n_1$, $n_2$-th Fourier coefficient is the Dirichlet series

$$R(\nu_1,\nu_2) = \varsigma(6\nu_1-1)\,\varsigma(6\nu_2-1)\,\varsigma(6\nu_1+6\nu_2-3)\sum_{C_1,C_2} H(C_1,C_2)NC_1^{-3\nu_1}NC_2^{-3\nu_2},$$

$$H(C_1,C_2) = \sum_{\substack{A_1,B_1 \bmod C_1\\ A_2,B_2 \bmod C_2\\ A_1,B_1,C_1 \text{ coprime}\\ A_2,B_2,C_2 \text{ coprime}\\ A_1C_2+B_1B_2+C_1A_2=0}} \kappa(\gamma)^2\, e\left(\frac{n_1B_1}{C_1}+\frac{n_2B_2}{C_2}\right),$$

where $\gamma$ is a matrix having the bottom row $(A_1,B_1,C_1)$, and whose involute has bottom row $(A_2,B_2,C_2)$. Incidentally, it is possible to realize this Dirichlet series also as the Mellin transform of the quadratic Eisenstein series on $GL(2)$, but the $GL(3)$ interpretation seems to give more information.

We shall only be concerned here with the case $n_1=n_2=1$. Then, assuming Patterson's conjecture, we may show that

$$\varsigma(2w)\sum|\tau(m)|^2\,Nm^{-w} = R(\tfrac{w}{3}+\tfrac{1}{6},\tfrac{w}{3}+\tfrac{1}{6}).$$

By (7), this identifies both Dirichlet series (4) and (6). This equation is consistent with the functional equations and the locations of the poles.

Now let us present an identity which is closely related to Patterson's conjecture, and which may be the key to both proving the conjecture, and to generalizing it. There exists a theta series $\theta_3$ on the 4-fold cover of $GL(3)$ with known Fourier coefficients. Restricting ourselves strictly to those coefficients parametrized by powers of a prime, it is sufficient to describe the coefficients $\tau(p^{k_1}, p^{k_2})$ where $0 \le k_1, k_2 < 4$, because of the periodicity theorem. The only nonvanishing such coefficients are

$$\tau(1,1) = 1, \qquad \tau(1,p) = \tau(p,1) = Np^{-1/2}\,\overline{g_1(1,p)},$$

$$\tau(p,p^2) = \tau(p^2,p) = \overline{g_1(1,p)}, \qquad \tau(p^2,p^2) = Np^{-1/2}\,\overline{g_1(1,p)}^{\,2}.$$

Now, denoting the theta function on the 4-fold cover of $GL(2)$ for definiteness as $\theta_2$, the *Rankin-Selberg convolution* of $\theta_2$ with $\overline{\theta}_3$ is the Dirichlet series

$$L(s, \theta_2 \times \overline{\theta}_3) = \sum_{m_1, m_2} \tau(m_1)\,\overline{\tau(m_1, m_2)}\,N(m_1 m_2^2)^{-s}.$$

It has a functional equation with respect to $s \mapsto 1 - s$. Patterson's conjecture is closely related to the formula

(8) $$L(s, \theta_2 \times \overline{\theta}_3) = \zeta(4s - \tfrac{3}{2})\,L(s, \overline{\theta}_2),$$

where $L(s, \overline{\theta}_2)$ is the "Mellin transform" $\sum \overline{\tau(m)}\,Nm^{-s}$.

In this form, there seems to be some hope of proving the conjecture, by a variation of the method originally used by Patterson [3] to determine the Fourier coefficients of the cubic theta function on $GL(2)$. For the left-hand side automatically has a functional equation by the theory of Rankin-Selberg convolutions. One might hope to show then by the converse theorem that there exists an automorphic form $\phi$ on the four-fold cover of $GL(2)$ such that

$$L(s, \theta_2 \times \overline{\theta}_3) = \zeta(4s - \tfrac{3}{2})\,L(s, \phi).$$

Then, by the method used by Patterson in [3] or otherwise, one would hope to show that $\phi = \overline{\theta}$.

Now let us propose a very general conjecture which includes both (7) and (8) as special cases. Let $\theta_r$ denote a theta function on the $n$-fold cover of $GL(r)$, where $n$ is fixed throughout the following discussion. If $r' \le r \le n - 1$, then we conjecture that

(9) $$L(s, \theta_r \times \overline{\theta}_{n-r'}) = \zeta\Big(ns - \tfrac{n-r+r'}{2}\Big)\,\zeta\Big(ns - \tfrac{n-r+r'+2}{2}\Big) \cdots \zeta\Big(ns - \tfrac{n+r-r'-2}{2}\Big)\,L(s, \theta_{r'} \times \overline{\theta}_{n-r}).$$

Moreover, we conjecture that (9) may be identified as the Fourier coefficient of an Eisenstein series of parabolic type $r, r'$ on $GL(r + r')$.

This conjecture is consistent with everything which we know, and seems almost certainly true, although we have no idea how it should be proved. It implies a great deal about the Fourier coefficients of the theta functions, and when the Fourier coefficients of the Eisenstein series are further investigated (work in progress in collaboration with Solomon Friedberg), we believe that a complete and satisfactory generalization of Patterson's conjecture will be at hand. What is lacking at this time is the generalization of (6) to the case where $r$ or $r'$ is greater than 2.

Finally, it should be mentioned that this conjecture has an analog for cusp forms. If $\phi$ is a cusp form on the $n$-fold cover of $GL(r)$, we conjecture that the Rankin-Selberg convolution $L(s, \phi \times \overline{\theta}_{n-r'})$ is equal to the Fourier coefficient of an Eisenstein series on the $n$-fold cover of $GL(r+r')$ involving $\phi$. Since the Fourier coefficients of $\theta_{n-1}$ are essentially known, if $r' = 1$, this conjecture is probably provable with the present state of knowledge. For this, it is not necessary to assume that $r \leq n-1$.

## References


1. D. Bump and J. Hoffstein, *On Shimura's Correspondence*, to appear in Duke Math. J.
2. D. Kazhdan and S. Patterson, *Metaplectic Forms*, Publ. Math. IHES No. 59 (1984).
3. S. Patterson, *A cubic analogue of the theta series*, J. reine angew. Math. **296** (1977).
4. S. Patterson, *The Fourier coefficients of biquadratic theta series*, Notes.
5. S. Patterson, *Whittaker models of generalized theta series*, in Sém. de th. des nombres (Paris), 1982-83, Birkhäuser (1984).
6. T. Suzuki, *Some results on the coefficients of the biquadratic theta series*, J. reine angew. Math. **340**.
7. T. Suzuki, *Global Whittaker models of generalized theta series*, Preprint (1986).

# Computational Problems in Arithmetic of Linear Differential Equations. Some Diophantine Applications

D.V. Chudnovsky, G.V. Chudnovsky*
Department of Mathematics
Columbia University
New York, New York 10027

**INTRODUCTION.** In diophantine approximation, particularly in effective methods of irrationality proving, a crucial role is played by sequences of integral (nearly integral) of rational or algebraic numbers satisfying finite order recurrences (with coefficients polynomial in indices). Generating functions of such sequences satisfy linear differential equations; and these generating functions belong to Siegel's class of $G$- functions [3]. In this presentation we examine $G$ - function and linear differential equations that can have $G$ - function solutions. We present a brief review of known results in this direction. Our new research is devoted mainly to the second order equations and three-term linear recurrences leading to classical Stieltjes' continued fraction expansions [50], [51]. A new instrument used in global and $p$-adic analysis of the linear differential equations is furnished by computational methods, including methods of computer algebra. Our numerical experiments, some of which are summarized in this presentation, show that linear differential equations satisfied by power series with nearly integral coefficients ($G$-function) possess special arithmetic properties that are associated with the existence of action of Frobenius. We present evidence for the conjecture that all such equations with $G$-function solutions arise from geometry, as Picard-Fuchs equations on variation of period (Hodge) structure of algebraic varieties. To study these, globally nilpotent equations, we use methods of diophantine approximations, particular Padé-type approximations.

To study linear differential equations $p$-adically, we examine "$p$-adic spectrum" of the Lamé equation. New examples of classes of globally nilpotent equations (of Lamé type) are constructed using Fuchsian arithmetic group (including quaternion groups). These equations are connected to diophantine equations to such numbers as $\zeta(2), \zeta(3)$ and periods of elliptic curves. Closely connected with these are new continued fraction expansions, presented below, generalizing Stieltjes-Rogers [49], [50], [51], [52], [53], [54] continued fraction expansions of integrals of elliptic functions. These new expansions are associated with Lamé equations with integral parameter $n$. We reexamine these equations, looking for those Lamé equations that have algebraic solutions only. The existence of such solutions is connected with zeros of special modular forms. We look,

*This work was supported in part by NSF grant DMS-8409626, Program O.C.R.E.A.E., U.S. Air Force (ACMP Program).

in particular, at Lamé equations over $\mathbf{Q}$, where the classification is, at least in principle, possible (and is achieved for small $n$).

Hypergeometric functions should, according to Siegel's conjecture [4] encompass the class of $G$-functions. Among these, special ${}_3F_2$ functions are associated with Ramanujan quadratic period relations [58], [48]. We present generalizations of these to other arithmetic triangle groups. We also build a new theory of $p$-adic "Ramanujan like" quadratic period relations that associate with every hypergeometric identity for an algebraic multiple of $1/\pi$ a series of mod$p^k$ congruences on partial sums of these series. Such congruences are naturally identifiable with action of Frobenius on linear differential equations on elliptic periods, specialized at curves with complex multiplication. Applications of these new congruences will be reported elsewhere.

Computational methods useful in theoretical study of elliptic curves, can also benefit from elliptic curve arithmetic. In the final part of this presentation we describe a class of fast interpolation algorithms generalizing FFT schemes. Such fast interpolation is particularly efficient in modular arithmetic on elliptic curves with highly composite Mordell-Weil group, because they can be used for an arbitrary prime $p$ without any special binary structure.

We thank the Computer Algebra Group of IBM for their interest, help and access to their system SCRATCHPAD.

# 1 Globally Nilpotent Linear Differential Equations.

In this section we look at arithmetic properties of solutions of linear differential equations, that have solutions with special arithmetic properties. These special arithmetic properties are the near integrality properties of power series expansions of functions, known as $G$-functions.

A class of $G$-functions was introduced by Siegel in [3], a paper more famous for its diophantine equation treatment and classical results on a related class of $E$-functions.

Siegel's study of $E$-functions [3-4], were significantly advanced since then by many researchers, see particularly [5-6]. We would like to mention in this connection that only relatively recently we have proved results on the best possible measure of diophantine approximations of values of $E$- functions at rational points [7]. These results present an ultimate effective version of the Schmidt theorem [8] for the values of $E$ - functions. The $E$ - function $f(x)$ are entire functions with power series expansion $f(x) = \sum_{n=0}^{\infty} a_n x^n/n!$, where $a_n \in \overline{\mathbf{Q}}$ and for every $\epsilon > 0$, we have $|a_n| < (n!)^\epsilon$, denom$\{a_0, \ldots, a_n\} \leq (n!)^\epsilon : n \geq n_1(\epsilon)$. In algebraic geometry and analysis, however, most of the interesting functions are analytic only in the finite part of the complex plane and have better $p$-adic convergence properties; these are $G$-functions.

<u>Definition 1.1.</u> A function $f(x)$ with the expansion at $x = 0$:

$$f(x) = \sum_{n=0}^{\infty} a_n x^n$$

is called a $G$-function, if $f(x)$ satisfies a linear differential equation over $\overline{\mathbf{Q}}(x)$, if coefficients $a_n$ are algebraic numbers, and if there is a constant $C > 1$ such that for all $n \geq 0$ the sizes of coefficients $a_n$ (i.e. the maximum of absolute values of $a_n$ and all its conjugates) and the common denominators den$\{a_0, \dots, a_n\}$ are bounded by $C^n$.

Siegel in [3] only outlined a program of $G$-function values irrationality and transcendence results, similar to his $E$- function theorems. Progress in this direction became dependent on additional global conditions, requiring that the $G$-function property be satisfied by all other solutions of the same differential equation. Such a property a $(G, C)$-condition, first formulated in [9], we used to prove some $G$-functions results [9], [12], [11], [13].

In [14] and [15] we had proved the general linear independence results for values of arbitrary $G$-functions at algebraic points (close to the origin), without any additional conditions. These results materialize Siegel's program after some 55 years. We also have proved the strong $(G, C)$-property for arbitrary $G$- functions [20]. This result, connected with our study of the Grothendieck conjecture, implies, e.g. that all previous results on $G$-function theory, proved under very restrictive conditions, are unconditionally valid for all $G$-functions. To describe our result, and the $(G, C)$-function conditions, one needs the definition of the $p$-curvature.

We consider a system of matrix first order linear differential equations over $\mathbf{Q}(x)$, satisfied by functions $f_i(x) : i = 1, \dots, n$:

$$df_i(x)/dx = \sum_{j=1}^{n} A_{i,j}(x) f_j(x), \tag{1.1}$$

for $A_{i,j}(x) \in \mathbf{Q}(x) : i, j = 1, \dots n$. Rewriting the system (1.1) in the matrix form

$$df^t/dx = Af^t; A \in M_n(\mathbf{Q}(x)),$$

one can introduce the $p$-curvature operators $\Psi_p$, associated with the system (1.1) following [16], [17]. The $p$- curvature operators $\Psi_p$ are defined for a prime $p$, as

$$\Psi_p = (d/dx - A)^p (\mathrm{mod} p).$$

Then $\Psi_p$ is a linear operator that can be represented as $\Psi_p = -A_p(\mathrm{mod} p)$, where one defines for $m \geq 0$,

$$(d/dx)^m \equiv A_m(\mathrm{mod}\, \mathbf{Q}(x)[d/dx](d/dx - A)). \tag{1.2}$$

Let $D(x)$ be a polynomial from $\mathbf{Z}[x]$ that is the denominator of $A$, i.e. $D(x)A_{i,j}(x)$ is a polynomial in $\mathbf{Z}[x]$ for $i, j = 1, \dots, n$. The *(G,C)-function condition* [10]- [11] of (1.1) means that (1.1) is satisfied by a system $(f_1(x), \dots, f_n(x))$ of $G$-functions, and that there exists a constant $C_2 > 1$, such that for any $N$, the common denominator of all coefficients of all polynomial entries of matrices $D(x)^m A_m(x)/m! : m = 0, \dots, N$, is growing not faster than $C_2^N$. With this conditions is closely related a *global nilpotence* condition [15-18] stating that the matrices $\Psi_p$ are nilpotent for almost all primes $p$. The $(G, C)$-condition implies the global nilpotence condition.

In [15] we proved the global nilpotence (and the $(G, C)$-function condition) of linear differential equations having a $G$-function solution. To prove this result we used Padé approximants of the second kind.

Theorem 1.2. Let $f_1(x), \ldots, f_n(x)$ be a system of $G$-functions, satisfying a system of first order linear differential equations (1.1) over $\overline{\mathbf{Q}}(x)$. If $f_1(x), \ldots f_n(x)$ are linearly independent over $\overline{\mathbf{Q}}(x)$, then the system (1.1) satisfies a $(G, C)$-function condition and is globally nilpotent. Any solution of (1.1) with algebraic coefficients in Taylor expansions is a $G$-function.

Padé approximation methods, used to prove the $G$-function Theorem 1.2, were also successfully applied by us to the study of the Grothendieck conjecture. The main tool in the study of the Grothendieck conjecture, and in the current study of globally nilpotent equations is the analytic method of Padé- (rational), and more general algebraic approximations to functions satisfying nontrivial complex analytic and arithmetic (p-adic) conditions. The corresponding group of results can be considered as a certain "local-global" principle. According to this principle, algebraicity of a function occurs whenever one has a near integrality of coefficients of power series expansion—*local conditions*, coupled with the assumptions of the analytic continuation (controlled growth) of an expanded function in the complex plane (or its Riemann surface)—a global, *archimedean* condition.

To prove the algebraicity of an integral expansion of an analytic function, assumptions on a uniformization of this function have to be made.

Our results from [19] and [20] were proved in the multidimensional case as well, to include the class of functions, uniformized by Jacobi's theta-functions (e.g. integrals of the third kind on an arbitrary Riemann surface). Moreover, our result includes "the nearly-integral" expansions, when the denominators grow slower than a typical factorial $n!$ denominator. In general, our results [19-20], show that $g+1$ functions in $g$ variables having nearly integral power series expansions at $\overline{x} = \overline{0}$ and uniformized near $\overline{x} = \overline{0}$ by meromorphic functions of finite order of growth ***are algebraically dependent.***

The first application of "local-global" principle was to the following:

The Grothendieck Conjecture. If a matrix system (1.1) of differential equations over $\overline{\mathbf{Q}}(x)$ has a zero $p$-curvature $\Psi_p = 0$ for almost all $p$, then this system (1.1) has algebraic function solutions only.

According to this conjecture, strong integrality properties of *all* power series expansions of solutions of a given linear differential equation imply that all these solutions are algebraic functions.

Methods of Padé approximations allowed us to solve the Grothendieck conjecture in important cases, [24], [15], [20] including the case of Lamé's equation, for integral $n$. In [25], it was shown that the Grothendieck conjecture is true for any linear differential equation all solutions of which can be parametrized by the meromorphic functions. The result was considerably generalized in [20] for equations, solutions of which can be parametrized by means of multidimensional theta-functions. To the class of these equations belong equations of rank one over arbitrary (finite) Riemann surfaces [20]:

Theorem 1.3 Any rank one linear differential equation over an algebraic curve, i.e. a first-order equation with algebraic function coefficients, satisfies the Grothendieck conjecture. Namely, if $\Gamma$ is an algebraic curve (given by the equation $Q(z, w) = 0$) over

$\overline{\mathbf{Q}}$, and if the rank one equation

$$\frac{dF}{F} = \omega(z,w)dz \tag{1.3}$$

over $\overline{\mathbf{Q}}(\Gamma)$ (for an Abelian differential $\omega dz$ on $\Gamma$) is globally nilpotent, then all solutions of (1.3) are algebraic functions.

The relationship of the $p$-curvature operators with the monodromy (Galois) group of a differential equation is extremely interesting. Our methods, involving various generalizations of Padé approximations, allow us to prove the Grothendieck conjecture for a larger class of differential equations, when additional information on a monodromy group is available. A technique from [27] (cf. [28]) using a random walk method, allowed us to treat crucially important class of equations $Ly = 0$, whose monodromy group is up to a conjugation a subgroup of $GL_n(\overline{\mathbf{Q}})$.

While the Grothendieck conjecture describes equations, *all* solutions of which have nearly integral expansions, it is more important to find out which equations possess nearly integral or $p$-adically overconvergent (i.e. convergent in the $p$-adic unit disc, or, at least, better convergent than the $p$-adic exponent) solutions.

The $p$-adic overconvergence and the nearly integrality of solutions hold for linear differential equation with a natural action of Frobenius. A class of equations, where the action of Frobenius was studied by Dwork, Katz, Deligne and others is the class of Picard-Fuchs differential equations (for variation of periods or homologies of smooth and singular varieties), see reviews [1-2].

Next, all evidence points towards the conjecture that the globally nilpotent equations are only those equations that are reducible to Picard-Fuchs equations (i.e. equations satisfied by Abelian integrals and their periods depending on a parameter). As Dwork puts this conjecture, all globally nilpotent equations come from geometry.

Our results on $G$-functions allow us to represent this conjecture even in a more fascinating form. We call this conjecture Dwork-Siegel's conjecture for reasons to be explained later:

Dwork-Siegel Conjecture. Let $y(x) = \sum_{N=0}^{\infty} c_N x^N$ be a $G$-function (i.e. the sizes of $c_N$ and the common denominators of $\{c_0, \dots, c_N\}$ grow not faster than the geometric progression in $N$). If $y(x)$ satisfies a linear differential equation over $\overline{\mathbf{Q}}(x)$ of order $n$ (but not of order $n-1$), then the corresponding equation is reducible to Picard-Fuchs equations. In this case $y(x)$ can be expressed in terms of multiple integrals of algebraic functions.

Siegel, in fact, put forward a conjecture which is, in a sense, stronger than the one given above. To formulate Siegel's conjecture we have to look again at his $E$-functions defined in [3]. Siegel showed that the class of $E$ - functions is a ring closed under differentiation and integration. Siegel also studied the hypergeometric functions

$${}_mF_n\left({a_1,\dots,a_m \atop b_1,\dots,b_n}\middle|\lambda x\right)$$

for algebraic $\lambda \neq 0$, *rational* parameters $a_1, \dots, a_m$ and $b_1, \dots, b_n$ and $m \leq n$. These functions he called hypergeometric $E$=functions and suggested in [4] all $E$-functions can be constructed from hypergeometric $E$-functions.

Looking at the (inverse) Laplace transform of $f(x)$, we see that Siegel's conjecture translates into a conjecture on $G$-function structure stronger than Dwork-Siegel's conjecture given above. Indeed, it would seem that all Picard-Fuchs equations might be expressed in terms of generalized hypergeometric functions.

This stronger conjecture is not entirely without merit; e.g. one can reduce linear differential equations over $\overline{\mathbf{Q}}(x)$, satisfied by $G$-functions to higher order equations over $\overline{\mathbf{Q}}(x)$ with regular singularities at $x = 0, 1, \infty$ only–(like the generalized hypergeometric ones) cf. [27].

We are unable so far to give a positive answer to this Dwork-Siegel conjecture, that all arithmetically interesting ($G-$)functions are solutions of Picard-Fuchs equations. Nevertheless, in some cases we can prove that this conjecture is correct. For now our efforts are limited to the second order equations (which provides with an extremely rich class of functions).

Proposition 1.4 Let a second order equation over $\mathbf{Q}(x)$: $Ly = 0$ be a globally nilpotent one and it has zero $p$-curvature $\Psi_p = 0$ for primes $p$ lying in the set of density $1/2$. Then the corresponding linear differential equation either have all of its solutions as algebraic functions, or is reducible to Picard-Fuchs equation (corresponding to the deformation of the curve), *or* has at least one transcendent element in a monodromy matrix for any representation of the monodromy group.

# 2 Arithmetic Properties and Diophantine Applications of Lamé Equations with n = -1/2.

Linear differential equations of the second order become arithmetically nontrivial when there is at least one accessory parameter. The first such case occurs for equations with 4 regular singularities. Among these a prominent role is played by the general Lamé equations; it is represented in the form [31-33]:

$$y'' + \frac{1}{2}\left\{\frac{1}{x} + \frac{1}{x-1} + \frac{1}{x-a}\right\}y' + \frac{B - n(n+1)x}{4x(x-1)(x-a)}y = 0 \tag{2.1}$$

depending on $n$ and on accessory parameters $B$.

A more familiar form of the Lamé equation is the transcendental one, with the change of variables: $a = k^{-2}$, $x = (sn(u,k))^2$, [31]:

$$\frac{d^2y}{du^2} + k^2 \cdot \{B - n(n+1)sn^2(u,k)\}y = 0 \tag{2.2}$$

in terms of Jacobi $sn$-function. An alternative form of (2.1-2) is in terms of Weierstrass' elliptic function:

$$\frac{d^2y}{du^2} + \{H - n(n+1)\mathcal{P}(u)\}y = 0. \tag{2.3}$$

Lamé equations are considered usually for integral values of the parameter $n$ in (2.1-3). This is the only case when solutions of (2.3) (or (2.2)) are meromorphic functions in the $u$-plane. In the case of integral $n$ the following facts are known [31,33]:

i) there exist $2n+1$ values of an accessory parameter ($B$ in (2.2) or $H$ in (2.3)) for which the algebraic form of the Lamé equation (2.1) has algebraic function solutions. These numbers $B_n^m : m = 1, \ldots, 2n+1$ are the ends of lacunas of the spectrum of an equation (2.2) considered as the spectral problem for the Lamé potential;

ii) all solutions of (2.2) and (2.3) are meromorphic functions of $u$ of order of growth 2.

Moreover, for every $B \neq B_n^m$, two linearly independent solutions of (2.3) have the form

$$F_{\pm} = \sqcap_{i=1}^{n} \frac{\sigma(a_i \pm u)}{\sigma(u)\sigma(a_i)} \cdot \exp\{\mp u \sum_{i=1}^{n} \varsigma(a_i)\}$$

for parameters $a_i$ determined from $B$—all $\wp(a_i)$ are algebraic in terms of $B$.

If the Lamé equation (2.1) is defined over $\overline{\mathbf{Q}}$ (i.e. $a \in \overline{\mathbf{Q}}$ and $B \in \overline{\mathbf{Q}}$) our local-global principle of algebraicity can easily solve the Grothendieck conjecture for Lamé equations with integral $n$. We have proved in [25]:

Theorem 2.1. For integer $n \geq 0$ the Lamé equation has zero $p$-curvature for almost all $p$ if and only if all its solutions are algebraic functions. The Lamé equation with integral $n$ is globally nilpotent for $2n+1$ values of $B$: $B = B_n^m$—ends of lacunas of spectrum of (2.3).

For all other values of $B$, the global nilpotence of the Lamé equations with integral $n$ over $\overline{\mathbf{Q}}$ is equivalent to the algebraicity of all solutions of (2.3).

The possibility of all algebraic solutions of (2.1) with $B \neq B_n^m$ was shown by Baldassari, and kindly communicated to us by Dwork. Such a possibility is discussed below.

For nonintegral $n$ no simple uniformization of solutions of Lamé equation exits. Moreover, Lamé equations themselves provide the key to several interesting uniformization problems. An outstanding Lamé equation is that with $n = -1/2$. This equation (and some of its equivalents to be seen later) determine the uniformization of the punctured tori. This leads to the classical Poincaré-Klein [34-35] problem of accessory parameter, which in the case of (2.1) with $n = -1/2$ means the determination for any $a \neq 0, 1, \infty$ a unique value of $B$, for which the monodromy group of (2.1) is represented by real $2 \times 2$ matrices. This complex-analytic investigation of the complex analytic structure of the Lamé (and of the more general) equation and the accessory parameter had been actively pursued by Klein, Poincaré, Hilbert, Hilb [39], V. I. Smirnov [36], Bers [37] and Keen [38]. Recently accessory parameter problem was studied in connection with conformally invariant field theories by Polyakov, Takhtajan, Zograf and others, cf. [40].

The uniformization problem for the punctured tori case is particularly easy to formulate, and our efforts towards the examination of the arithmetic nature of Fuchsian groups uniformizing algebraic curves were initially focused on this case. The punctured tori case can be easily described in terms of Lamé equation with $n = -1/2$. If one starts with a tori corresponding to an elliptic curve $y^2 = P_3(x)$, then the function inverse to the automorphic function, uniformizing the tori arises from the ratio of two solutions of the Lamé equation with $n = -1/2$. If $P_3(x) = x(x-1)(x-a)$ (i.e. the singularities are at $x = 0$, 1, $a$ and $\infty$), then the monodromy group of (2.1) is determined by 3 traces

$x = \mathrm{tr}(M_0 M_1)$, $y = \mathrm{tr}(M_0 M_a)$, $z = \mathrm{tr}(M_1 M_a)$. Here $M$ is a monodromy matrix in a fixed basis corresponding to a simple loop around the singularity $a$. These traces satisfy a single Fricke identity [38]:

$$x^2 + y^2 + z^2 - xyz = 0.$$

There exists a single value of the accessory parameter $C$ for which the uniformization takes place. Equivalently, $C$ is determined by conditions of reality of $x, y, z$.

Algebraicity Problem [32]. Let an elliptic curve be defined over $\overline{\mathbf{Q}}$ (i.e. $a \in \overline{\mathbf{Q}}$). Is it true that the corresponding (uniformizing) accessory parameter $C$ is algebraic? Is the corresponding Fuchsian group a subgroup of $GL_2(\overline{\mathbf{Q}})$ (i.e. $x, y$ and $z$ are algebraic)?

Extensive multiprecision computations, we first reported in [32], of accessory parameters showed rather bleak prospect for algebraicity in the accessory parameter problems. Namely, as it emerged, there are only 4 (classes of isomorphisms of) elliptic curves defined over $\overline{\mathbf{Q}}$, for which the values of uniformizing accessory parameters are algebraic. These 4 classes of algebraic curves are displayed below in view of their arithmetic importance.

Why are we interested in algebraicity (rationality) of the accessory parameters?

It seems that attention to the arithmetic properties of the Lamé equation with $n = -1/2$ arose shortly after Apéry's proof of the irrationality of $\varsigma(2)$ and $\varsigma(3)$. His proof (1978), see [42], was soon translated into assertions of integrality of power series expansions of certain linear differential equations.

To look at these differential equations we will make use of the classical equivalence between the punctured tori problem and that of 4 punctures on the Riemann sphere. For differential equations this means Halphen's algebraic transformation from [31], [41] between the Lamé equation with $n = -1/2$:

$$P(x)y'' + \frac{1}{2}P'(x)y' + \frac{x + C}{16}y = 0, \tag{2.4}$$

for $P(x) = x(x-1)(x-a)$, and the Heun equation with zero-differences of exponents at all singularities:

$$P(x)y'' + P'(x)y' + (x + H)y = 0. \tag{2.5}$$

The relation between two accessory parameters is the following

$$C = 4H + (1 + a).$$

Let us denote the equation (2.5) by $Ly = 0$. We have already stated that there are 4 Lamé equations with $n = -1/2$ (up to Möbius transformations) for which the value of the accessory parameter is known explicitly and is algebraic. These are 4 cases when the Fricke equation

$$x^2 + y^2 + z^2 = xyz,$$

with $0 \le x \le y \le z \le xy/2$, has solutions, whose squares are integers. It is in these 4 cases, when the corresponding Fuchsian group uniformizing the punctured tori is the congruence (arithmetic) subgroup, see references in [32], [27].

Let us look at these 4 cases, writing down the corresponding equation (2.5):
1) $x(x^2-1)y''+(3x^2-1)y'+xy=0$.
2) $x(x^2+3x+3)y''+(3x^2+6x+3)y'+(x+1)y=0$.
3) $x(x-1)(x+8)y''+(3x^2-14x-8)y'+(x+2)y=0$.
4) $x(x^2+11x-1)y''+(3x^2+22x-1)y'+(x+3)y=0$.
Each of the equations 1)-4) is a pull-back of a hypergeometric function by a rational map, and, as a consequence, each of the equations is globally nilpotent. In fact, for each of the equations 1)-4) there exist an integral power series $y(x)=\sum_{N=0}^{\infty} c_N x^N$ satisfying $Ly=0$ and regular at $x=0$.

Apéry's example for $\varsigma(2)$ arises from the equation 4). In this case the solution $y(x)$ of 4), regular at $x=0$ has the form $y(x)\text{-}\sum_{N=0}^{\infty} c_N x^N$ with integral $c_N$. It is easy to see that a nonhomogeneous equation $Ly=\text{const}\neq 0$ has a solution $z(x)=\sum_{N=0}^{\infty} d_N x^N$ regular at $x=0$ with nearly integral $d_N$ (this is according to the global nilpotence of the corresponding $L$).

The only (possible) singularities of $y(x)$ and $z(x)$ in the finite part of the plane are $x=(-11\pm 5\sqrt{5})/2$, where all local exponents are zero. Thus we can always find a constant $\varsigma$ such that $\varsigma\cdot y(x)+z(x)$ is regular at $x=\frac{-11+5\sqrt{5}}{2}=(\frac{\sqrt{5}-1}{2})^5$. Apéry determined theconstant $\varsigma$; if one takes $z(x)=\sum_{N=0}^{\infty} d_N x^N$ as $Lz=5$, then $\varsigma=\varsigma(2)=\pi^2/6$.

One obtains the irrationality of $\varsigma(2)$, because $c_N/d_N$ are dense approximations, and a nontrivial measure of irrationality of $\pi^2$ is derived too.

Other equations 1)-3) can be used in a similar way, because of their global nilpotence and existence of only 4 regular singularities. Apéry approximations to $\varsigma(3)$ arises from the equation 3), or rather from a symmetric square of the corresponding operator $L$.

These examples lead to a method of the construction of sequences of dense approximation to numbers using nearly integral solutions of globally nilpotent equations. Often the corresponding equations are Picard-Fuchs equations satisfied by generating functions of Padé approximants to solutions of special linear differential equations, see examples in [43], [44]. Apéry's approach is not the best for the purpose of improving measures of irrationality. But it gives a good starting point.

Diophantine approximations suggest the following problem: determine all cases of global nilpotence of Lamé equations.

Our intensive numerical experiments reveal a predictable, phenomenon: it seems that, with the exception of equations 1)-4) (and equations equivalent to them via Möbius transformations), there is *no* Lamé equations with $n=-1/2$ over $\overline{\mathbf{Q}}$ that are globally nilpotent. We put these observations as a

Conjecture 2.2. Lamé equations with $n=-1/2$ defined over $\overline{\mathbf{Q}}$ are not globally nilpotent except for 4 classes of equations corresponding to the congruence subgroups, with representatives of each class given by 1)-4).

What are our grounds for this conjecture? First of all, Padé approximation technique related to the Dwork-Siegel conjecture allows us to prove one positive result in the direction of this conjecture for the $n=-1/2$ case of the Lamé equation.

Theorem 2.3. For fixed $a \in \overline{\mathbf{Q}}$ ($a \neq 0,1$), there are only finitely many algebraic numbers $C$ of bounded degree $d$ such that the Lamé equation with $n = -1/2$ is globally nilpotent.

Similar results hold for any Lamé equation with rational $n$.

Of course, one wants a more specific answer for any $n$ (e.g. for $n = -1/2$, there are only 4 classes of $a$ and $C$ given above with the global nilpotence). However for half-integral $n$, there are always $n + 1/2$ trivial cases of global nilpotence, where solutions are expressed in terms of elliptic integrals, see [31], [41].

We have started the study of globally nilpotent Lamé equations (2.4) or (2.5) with numerical experiments. This ultimately led to Conjecture 2.2.

We checked possible equations of the form (2.5) with

$$P(x) = x(x^2 - a_1x + a),$$

i.e. 4 singularities at $x = 0, \infty$ and 2 other points, for values of

$$a_1, a \in \mathbf{Z}$$

in the box: $|a|, |a_1| \leq 200$. For all these equations (2.5) we checked their $p$-curvature for the first 500 primes. Our results clearly show that with an exception of 4 classes of equations equivalent to 1)-4), any other equation has a large proportion of primes $p$ such that the $p$-curvature is not nilpotent for any $H \in \mathbf{Q}$!

An interesting $p$-adic problem arises when, instead of globally nilpotent equations one looks at the nilpotence conditions of $p$ curvature for a fixed $p$ or, equivalently, when thee is a $p$- integral solution for a fixed $p$. The only known case (Tate- Deligne) corresponds to Lamé equation with $n = 0$, where the unit root of the $\varsigma$-function of the corresponding elliptic curve is expressed in terms of a unique accessory parameter. This example suggests a definition of $p$-adic spectrum, which we use only for Lamé equations.

We are interested in those $H$ mod $p$ for which the $p$-curvature of (2.6) is nilpotent, and particularly in those $p$- adic $H \in \mathbf{Z}_p$ for which there exists a solution $y = y(x)$ of (2.6) whose expansion has $p$-integral coefficients. We call those $H \in \mathbf{Z}_p$, for which such $y(x)$ exists, eigenvalues of (2.6) in the "$p$-adic domain", and their set we call "an integral $p$-adic spectrum". The problem of study of arithmetic nature of Lamé equation was proposed by I.M. Gelfand. Of course, if a $p$-curvature is not zero, there is no second solution $y(x)$ with the same property.

To determine $p$-adic spectrum we conducted intensive symbolic and numerical computations using SCRATCHPAD (IBM) and MACSYMA (Symbolics Inc.) systems.

We start with the observations of the "mod $p$" spectrum as $p$ varies.

I. For noncongruence equations (2.5) with rational $a \neq 0, 1$ (i.e. for an elliptic curve defined over $\mathbf{Q}$ with a point of order 2) there seem always to be infinitely many primes $p$ for which *no* value of the accessory parameter $H$ mod $p$ gives a nilpotent $p$-curvature (thus mod $p$ spectrum is empty).

Here are a few statistics for noncongruence equations with rational integers $a$:

For $a = 3$ the first $p$'s with the null spectrum mod $p$ are: $p =$ 61, 311, 677, 1699, 1783, 1811, 2579, 2659, 3253,... .

For $a = 5$ the first $p$'s with the null spectrum mod $p$ are: $p = 659$, 709, 1109, 1171, 1429, 2539, 2953, 2969, 3019, 3499, 3533, 3803, 3863, 4273, 4493, 4703, 4903, 5279, 5477, 5591, 6011, 7193, 7457, 7583,... .

For $a = 4$ the corresponding $p$'s with the null spectrum are: $p = 101$, 823, 1583, 2003, 3499,... .

For $a = 13$ the corresponding list starts at: $p = 1451$, 1487, 2381,... .

Observation I above was checked by us for all noncongruent $P(x) = x(x^2 - a_1x + a)$ with integral $a_1, a$ not exceeding 250 in absolute value.

II. An integral $p$-adic spectrum of equations (7.2) with ($p$- integral) $a$ has a complicated structure depending on the curve.

$p$-adic spectrum can be null, finite (typically a single element), or infinite, resembling the Cantor set.

Numerical analysis is not easy either. To study $p$-adic expansions of $p$-adic numbers from $p$-adic spectrum up to the order mod$p^k$, one has to carry all the computations of rational functions and coefficients of power series expansions in the modular arithmetic mod$p^N$ for $N = 2(p^k - 1)/(p - 1)$. For example, in order to determine the 3-adic spectrum with 14 digits of precision (in the 3-adic expansion), one has to carry out computations with over 2,000,000 decimal digit long numbers! For congruence equations 1)-4) the $p$-adic spectrum seems to be an infinite one with a Cantor-like structure.

Like with the complex-analytic uniformization problem, there is a relationship with the $p$-adic uniformization of Mumford, particularly for curves with the multiplicative reduction at $p$.

# 3 Arithmetic of Lamé Equation for Different n.

Examples above show how globally nilpotent equations can be used for various irrationality and transcendence proofs. Picard- Fuchs equations, for example, provide generating functions for Padé approximants in Padé approximation problem to generalized hypergeometric functions, Pochhammer integrals and other classes of solutions of linear differential equations. We suggest, as a starting equation, when it is of the second order, an equation corresponding to an arithmetic Fuchsian subgroup. Congruence subgroups of $\Gamma(1)$ and quaternion groups provide with interesting families of globally nilpotent equations.

One can start with equations uniformizing punctured tori with more than one puncture. The complete description of arithmetic Fuchsian groups of signature $(1; e)$ had been provided by Maclachlan and Rosenberger [46] and Takeuchi [47]. These groups of signature $(1; e)$ are defined according to the following representation

$$\Gamma = <\alpha, \beta, \gamma | \alpha\beta\alpha^{-1}\beta^{-1}\gamma = -1_2, \gamma^e = -1_2>,$$

where $\alpha$ and $\beta$ are hyperbolic elements of $SL_2(\mathbf{R})$ and $\gamma$ is an elliptic (respectively a parabolic) element such that $\mathrm{tr}(\gamma) = 2\cos(\pi/e)$.

For all $(1; e)$ arithmetic subgroups there exists a corresponding Lamé equation with a rational $n$, uniformized by the corresponding arithmetic subgroup. This way we obtain *73* Lamé equations, all defined over $\overline{\mathbf{Q}}$ (i.e. the corresponding elliptic curves and accessory parameter $C$ are defined over $\overline{\mathbf{Q}}$). Some of these equations give rise to nearly integral sequences satisfying three-term linear recurrences with coefficients that are quadratic polynomials in $n$, and have the growth of their denominators and the convergence rate sufficient to provide the irrationality of numbers arising in this situation in a way similar to that of Apéry.

Groups of the signature $(1;e)$ correspond to the Lamé equations (see (2.1)):

$$P(x)y'' + \frac{1}{2}P'(x)y' + \{C - \frac{n(n+1)}{4}x\}y = 0$$

with $n + \frac{1}{2} = \frac{1}{2e}$.

In the arithmetic case one looks at totally real solutions of the modified Fricke's identity, which now takes the form:

$$x^2 + y^2 + z^2 - xyz = 2(1 - \cos(\frac{\pi}{e})).$$

Using numerical solution of the (inverse) uniformization problem, we determined the values of the accessory parameters. Among the interesting cases are the following:

Here $P(x) = x(x-1)(x-A)$ and:

(1;2)-case:

1) $A = 1/2$, $C = -3/128$,

$$(x = y = (1+\sqrt{2})^{\frac{1}{2}} \cdot 2^{\frac{3}{4}}, z = 2+\sqrt{2});$$

2) $A = 1/4$, $C = -1/64$;
3) $A = 3/128$, $C = -13/2^{11}$;
4) $A = (2-\sqrt{5})^2$, $C = \sqrt{5} \cdot (2-\sqrt{5})/64$;
5) $A = (2-\sqrt{3})^4$, $C = -(2-\sqrt{3})^2/2^4$;
6) $A = (21\sqrt{33} - 27)/256$.

(1;3)-case

1) $A = 1/2$, $C = -1/36$
2) $A = 32/81$, $C = -31/2^4 \cdot 3^4$;
3) $A = 5/32$, $C = -67/2^9 \cdot 3^2$;
4) $A = 1/81$, $C = -1/2 \cdot 3^4$;
5) $A = (8 - 3\sqrt{7})/2^4$.

(1;4)-case

1) $A = -11 + 8\sqrt{2}$, $C$-cubic;
2. $A = (3 - \sqrt{8})/4$, $C$-cubic;

(1;5)-case

1) $A = 3/128$, $C = -397/2^{11} \cdot 5^2$;

In all cases above, $A$ is real (as well as $C$) and $0 < A \le 1/2$.

Not all elliptic curves corresponding to $(1; e)$-groups are defined over $\mathbf{Q}$—there is a nontrivial action of the Galois group (cf. with a different situation in [47]).

Let us return to the case of integral $n$, to complete the classification problem started in Theorem 2.1.

For linear differential equations, whose solutions are parametrized by meromorphic functions, our local-global algebraicity principle [19], [20], [25] proves the Grothendieck conjecture: if $p$-curvature is zero for almost all (density one) primes $p$, then all solutions of the equation are algebraic functions. A class of such linear differential equations includes equations known as "finite-band potentials" (familiar from the Korteweg-de Vries theory), among which the most prominent are Lamé equations with integer parameter $n$. These equations are parametrized by Baker's functions that are solutions of rank one equations over curves of positive genus, see [19]. For rank one equations the Grothendieck conjecture was proved in [19], see Theorem 1.3. According to this solution, if a rank one equation is globally nilpotent, its solutions are algebraic functions. Particularly simple and self contained proofs in the elliptic curve case can be found in §6 of [19] and in [25]. For Lamé equations our result implies that Lamé equations over $\overline{\mathbf{Q}}(x)$ with integral $n$ can be globally nilpotent if and only if either the accessory parameter $B$ is one of the lacunas ends $B_n^m$ ($m = 1, \ldots, 2n + 1$), or else all solutions of Lamé equations are algebraic functions. In [25] we implicitly assume that Lamé equations over $\overline{\mathbf{Q}}(x)$ with integral $n$ cannot have a finite monodromy group (all solutions are algebraic functions). In fact, Lamé equations for integral $n \ge 1$ with algebraic solutions only are possible. These exceptional equations correspond to nontrivial zeros of special modular forms, and are closely connected with interesting algebraicity problems for exponents of periods of incomplete elliptic integrals of the third kind. To understand these relationships we use Hermite's solution of Lamé equations [31]. Let us look at Weierstrass elliptic functions $\sigma(u), \varsigma(u)$ and $\mathcal{P}(u)$, associated with the lattice $L = 2\omega_1 \mathbf{Z} + 2\omega_2 \mathbf{Z}$ in $\mathbf{C}$. The Hermite's function is

$$H(v; u) = \frac{\sigma(u - v)}{\sigma(u)\sigma(v)} e^{\varsigma(v)u}. \tag{3.1}$$

For $n = 1$ and $B = \mathcal{P}(v)$ ($\ne B_1^m$), two linearly independent solutions of Lamé equations are $H(\pm v; u)$. If $n \ge 1$ and $B \ne B_n^m$ ($m = 1, \ldots, 2n + 1$) then two linearly independent solutions of the Lamé equation with an integral $n$ parameter can be expressed in the form

$$F = \sum_{j=0}^{n-1} b_j \frac{d^j}{du^j} \{H(v; u) \cdot e^{\rho u}\},$$

where $b_0, b_1, \ldots b_{n-1}, \rho, \mathcal{P}(v)$ are determined algebraically over $\mathbf{Q}$ in terms of $B$ and parameters $g_2, g_3$ of $\mathcal{P}(x)$.

The monodromy group of Lamé equations with integral $n$ can be thus easily expressed explicitly in terms of Floquet parameters. The general theory is outlined in [41] for arbitrary equations of "Picard type", where Floquet solutions in this, doubly periodic case, are called multiplicative solutions. Whenever the accessory parameter $B$ is distinct from lacunas ends, the monodromy can be determined from the action of two translations $u \to u + 2\omega_1$ or $u \to u + 2\omega_2$ on Hermite's function $H(v, u)$. The rule of transformation of $H(v; u)$ is very simple: if $2\omega = 2m_1\omega_1 + 2m_2\omega_2$, then

$$H(v; u + 2\omega) = H(v; u) : \exp\{\varsigma(v)2\omega - v \cdot 2\eta\},$$

for $2\eta = 2m_1\eta_1 + 2m_2\eta_2$. Thus with every $H(v; u)$ function two Floquet parameters $\mu_1$ and $\mu_2$ are associated:

$$\mu_i = \exp\{\varsigma(v) \cdot 2\omega_i - v \cdot 2\eta_i\} : i = 1, 2. \tag{3.2}$$

Let us look at Floquet solutions of an arbitrary Lamé equation with integral $n$ (or following notations of [41] at an arbitrary multiplicative solution of Picard equation). Such a solution can be expressed in the form

$$F(u) = H(v; u)e^{\rho u} \cdot P(u), \tag{3.3}$$

where $P(u)$ is an elliptic function of $u$. For such a solution to be algebraic in $\wp(u)$ it is necessary and sufficient for its Floquet parameters to be roots of unity. The Floquet parameters $s_i$ of $F(u)$ defined as

$$F(u + 2\omega_i) = F(u) \cdot s_i : i = 1, 2,$$

are $s_i = \mu_i \cdot e^{\rho 2\omega_i} =$ ~~$\exp\{\varsigma(v) \cdot v \cdot 2\eta_i + 2\rho\omega_i\}$~~ $\exp\{\zeta(v)2\omega_i - v 2\eta_i + 2\rho\omega_i\}$

This gives two equations on $v$, solving which and taking into the account the Legendre identity $\eta_1\omega_2 - \eta_2\omega_1 = \frac{1}{2}\pi\sqrt{-1}$, we get

$$\varsigma(v) + \rho = -r_1 \cdot 2\eta_2 + r_2 \cdot 2\eta_1;$$

$$v = -r_1 \cdot 2\omega_2 + r_2 \cdot 2\omega_1,$$

if $s_i = \exp\{r_i \cdot 2\pi\sqrt{-1}\}$ for rational $r_i \in \mathbb{Q} : i = 1, 2$. This shows that $F(u)$ can be an algebraic function in $\wp(u)$ only if (but not necessarily) if $v$ is a torsion point of $L$. The precise relations are presented above:

$$v = \frac{n_1 \cdot 2\omega_1 + n_2 \cdot 2\omega_2}{N};$$

$$\varsigma(v) + \rho = \frac{n_1 \cdot 2\eta_1 + n_2 \cdot 2\eta_2}{N}, \tag{3.4}$$

where $(n_1, n_2, N) = 1$. These relations express conditions on an elliptic curve (depending on $N$ and the dependency of $\rho$ on $v$) to have a solution $F(u)$ algebraic in $\wp(u)$ To express this relation in a more explicit form we use multiplication formula for elliptic functions. One of the best expression of multiplication formulas involves elliptic functions

$$\psi_N(u) \stackrel{def}{=} \frac{\sigma(Nu)}{\sigma(u)^{N^2}}$$

for $N \geq 1$. These elliptic functions satisfy the famous three-term (Weierstrass) nonlinear recurrences, and some properties of these functions and their specializations are summarized in [74]. Using the $\psi_N(u)$-polynomials in $\wp(u)$ and $\wp'(u)$, we derive a multiplication formula for $\zeta(u)$:

$$\zeta(Nu) - N\zeta(u) = \frac{1}{N}\frac{\frac{\partial}{\partial u}\psi_N(u)}{\psi_N(u)},$$

representing $\zeta(Nu) - N\zeta(u)$ as a rational function in $\wp(u), \wp'(u)$. This multiplication formula cannot, unfortunately be used directly for $u = v$ - the torsion point of order $N$ exactly, if $N$ is odd, because $\zeta(u)$ has poles at lattice points. We can modify it, though, by considering $u = x + v$ at $x \to 0$ for $N$-th order torsion point $v$. This way we get for $v = (n_1 \cdot 2\omega_1 + n_2 \cdot 2\omega_2)/N$,

$$\{(n_1 \cdot 2\eta_1 + n_2 \cdot 2\eta_2)/N - N \cdot \zeta(v)\} = \frac{1}{2N} \cdot \frac{\psi_N''(v)}{\psi_N'(v)}.$$

These multiplication laws allow us to express the conditions on $F(u)$ to be an algebraic function in $\wp(u)$ in a concise form:

$$\psi_N(v) = 0 (\text{or } v \in \frac{1}{N}L \backslash L);$$

$$\rho = \frac{1}{2N} \cdot \frac{\psi_N''(v)}{\psi_N'(v)}. \tag{3.5}$$

[The last expression always makes sense, because $\sigma(u)$ does not have multiple roots.]

This system of equations for a given $N$ is actually a single condition on the parameter $\tau = \omega_1/\omega_2$ in $H$ of the elliptic curve. Such conditions and explicit expressions of $\rho$ and $B$ in terms of $\wp(v)$, allow us to find, for a given $N$ and $n$, all the Lamé equations with integral parameter $n$ that have algebraic solutions only (with the order of local monodromy group dividing $N$). For a given $n$ this seems to give an infinite set of conditions (parametrized by $N$). In reality, at least for a fixed field of definition of the Lamé equation, the determination of all cases of algebraicity is easier. Let us look at those Lamé equations that are defined over $\mathbf{Q}(x)$, i.e. $g_2, g_3$ and $B$ are rational numbers. We are looking at those torsion points $v$ of order $N$ that are defined over $\mathbf{Q}$ ($\wp(v) \in \mathbf{Q}$). In view of Masur's theorem, $N$ is bounded. This leaves us for any given $n$ only with finitely many cases corresponding to $N = 2,3,4,5,6,7,8,9,10,12$. Nonobvious generalizations of this argument are possible for arbitrary algebraic number field $K$ and Lamé equations defined over $K(x)$. Elliptic curves determined by such conditions have parameters $\tau$ that are nontrivial zeros of special modular forms of weights depending on $n$ and $N$. For any given $n$ and $N$ such nontrivial zeros, and thus the invariants $g_2$ and $g_3$ of the corresponding elliptic curves, can be explicitly determined.

Let us take $n = 1$ in the Lamé equation (the case $n = 0$ is trivial—only $B = 0$ gives a globally nilpotent equation). For $n = 1$ and $B = \wp(v)$ ($\neq B_1^m = e_m = \wp(\omega_m)$), two linearly independent solutions of the Lamé equation are $H(\pm v; u)$, i.e. $\rho = 0$ in the expression for $F(u)$ above. Consequently, for $n = 1$ and a fixed $N$ the equations defining the elliptic curve and $v$ are:

$$\psi_N(v) = 0, \ \psi_N''(v) = 0. \tag{3.6}$$

This immediately shows that the case of even $N$ is impossible. For odd $N \leq 9$ all solutions can be easily found using any computer algebra systems. We summarize these findings choosing for an elliptic curve simple notations:

$$y^2 = x^3 + ax + b,$$

i.e. $a = -g_2/4$, $b = -g_3/4$. For $N = 3$ the only solution of the equations above is

$$a = 0, \ b \neq 0 - \text{arbitrary}, \tag{3.7}$$

i.e. $y^2 = x^3 + 1$ is the only exceptional curve. For $N = 5$ one gets an equation on $a, b$:

$$2160b^2 - 6241a^3 = 0, \tag{3.8}$$

or $a = 2^2 \cdot 3 \cdot 5$, $b = 2 \cdot 5 \cdot 79$ is the solution with $a, b \in \mathbf{Z}$ and the corresponding torsion point $x$ $(= \mathcal{P}(v))$ is $x = -4$.

For $N = 7$ or $N = 9$ there is no rational solutions in $a$ and $b$. For example, for $N = 7$ one gets the following equation on $a$ and $b$ that leads to a quadratic equation on the absolute invariant $J$:

$$-106709177088b^4 + 73256324400a^3b^2 + 137751312727a^6 = 0.$$

In short, there are only two Lamé equations with $n = 1$ over $\mathbf{Q}(x)$ having algebraic solutions only. They correspond to $N = 3$ and $N = 5$ and to two elliptic curves (3.6), (3.7) over $\mathbf{Q}$ given above.

The problem of algebraic solutions of Lamé equation has interesting transcendental translations. As we have seen, we are looking at Floquet parameters $\mu_i$ depending on $v$ such that $\mathcal{P}(v) \in \overline{\mathbf{Q}}$. One can ask a more general question: is $\mu_i$ algebraic (a root of unity)? As it was reported in [33], the transcendence theory shows that each $\mu_i = \exp\{\varsigma(v) \cdot 2\omega_i - v \cdot 2\eta_i\}$ is *transcendental*, whenever $v$ is not a torsion point, $\mathcal{P}(v) \in \overline{\mathbf{Q}}$ and $g_2, g_3 \in \overline{\mathbf{Q}}$. Nothing had been known so far about torsion points $v$. To present a definitive result in this direction, we start with a reformulation of one case of Schneider-Lang theorem. If an elliptic curve, corresponding to $\mathcal{P}$, $\mathcal{P}$'s is defined over $\overline{\mathbf{Q}}$, if $\mathcal{P}(v) \in \overline{\mathbf{Q}}$, $\rho \in \overline{\mathbf{Q}}$ and, as before in (3.3), $F(u) = H(v; u)e^{\rho u}$, then for a nonalgebraic $F(u)$, its Floquet parameter $F(u + \omega)/F(u)$ for $\omega \in L$, $\omega \neq 0$, is a transcendental number. On the other hand, if $F(u)$ is algebraic, then all its Floquet parameters are roots of unity (and thus algebraic). The classification problem for Lamé equations with algebraic solutions only is thus equivalent to the determination of all algebraic numbers of the form $\mu_i$ above. For elliptic curves over $\mathbf{Q}$ our results show that there are only two (up to isomorphism) elliptic curves with $\mathcal{P}(v) \in \mathbf{Q}$ such that $\mu_i$ is algebraic (third and fifth roots of unity).

# 4 Arithmetic Continued Fractions.

The problem of explicit determination of all linear differential equations that have arithmetic sense (i.e. an overconvergence property or the existence of nontrivial solutions mod $p$) can be easily translated into a classical problem of nearly integral solutions to linear recurrences. This problem arose in works of Euler, Lambert, Lagrange, Hermite, Hurwitz, Stieltjes and others in connection with irrational continued fraction expansions of classical functions and constants.

Problem 4.1. Let $u_n$ be a solution of a linear recurrence of rank $r$ with coefficients that are rational (polynomial) in $n$:

$$u_{n+r} = \sum_{k=0}^{r-1} A_k(n) \cdot u_{n+k}$$

for $A_k(n) \in \overline{\mathbf{Q}}(n)$ : $k = 0, \ldots, r-1$, and such that $u_n$ are "nearly integral". Then the generating function of $u_n$ is a function whose local expansion represents either an integral of an algebraic function or a period of an algebraic integral, i.e. a solution of Picard- Fuchs-like equation. The "near integrality" of $u_n$ means that $u_n$ are algebraic numbers whose sizes grow slower than factorials, i.e. for any $\epsilon > 0$, the sizes of $u_n$ are bounded by $(n!)^\epsilon$, and whose common denominator also grows slower than a factorial: i.e. for any $\epsilon > 0$ the common denominator of $\{u_0, \ldots, u_n\}$ is bounded by $(n!)^\epsilon$.

For continued fractions this problem can be reformulated:

Problem$'$. Let us look at an explicit continued fraction expansion with partial fractions being rational functions of indices:

$$\alpha = [a_0; a_1, \ldots, A(n), A(n+1), \ldots],$$

for $A(n) \in \mathbf{Q}(n)$. Let us look then at the approximations $P_n/Q_n$ to $\alpha$ defined by this continued fraction expansion:

$$\frac{P_n}{Q_n} = [a_0; a_1, \ldots, A(n-1), A(n)] :$$

$n \geq 1$, where $P_n, Q_n \in \mathbf{Z}$.

If the continued fraction representing $\alpha$ is convergent *and* for *some* $\epsilon > 0$

$$|\alpha - \frac{P_n}{Q_n}| < |Q_n|^{-1-\epsilon} :$$

$n \geq n0(\epsilon)$, i.e. if $\alpha$ is *irrational*, then the sequences $P_n$ and $Q_n$ of numerators and denominators in the approximations to $\alpha$ are arithmetically defined sequences; their generating functions represent solutions of Picard-Fuchs and generalized Picard-Fuchs equations.

The later equations correspond to deformations with possible irregular singularities, arising from Laplace and Borel transforms of solutions of ordinary Picard-Fuchs equations.

*Remark.* While partial fractions $a_n = A(n)$ are rational functions of $n$, the sequences $P_n$ and $Q_n$ are *not* rational or algebraic functions of $n$ unless in very special cases, when $\alpha$ is reducible to a rational number.

One of the main purposes of our investigation was an attempt to establish, first empirically, that there are only finitely many classes of such continued fraction expansions all of which can be determined explicitly. One has to distinguish several types of numbers/functions $\alpha$ and Picard-Fuchs like equations that can occur when such a continued fraction expansion of $\alpha$ exists:

A. $\Theta$-function parametrization. This is the case when a linear differential equation can be parametrized by Abelian or $\theta$-functions. This is the case of linear differential equations reducible to the so-called finite band/isospectral deformation equations. In general, the continued fraction expansions representing appropriate $\alpha$ do not have an arithmetic sense. Here $\alpha$ depends on the spectral parameter (uniformizing parameter of the curve) and on the curve moduli. For special values of spectral parameter ("ends of lacunas"), $\alpha$ is represented as a convergent continued fraction expansion with an arithmetic sense. In this case we had completely determined all the cases of global nilpotence in our work on the Grothendieck conjecture, see [19]. We return to the class A of continued fractions in connection with Stieltjes-Rogers continued fraction expansions.

B. In this case the monodromy group of a linear differential equation associated with a linear recurrence of any rank is connected with one of triangle groups. These groups do not have to be arithmetic. The cases of finite Schwarz's groups and elliptic groups are easier to describe. The hyperbolic (Fuchsian) cases provide with a large class of equations of high rank that are the blowups of hypergeometric equations. This is the case of Apéry's recurrences and continued fractions. However, for any *given* rank $r$ there are only *finitely many* linear differential equations that occur this way.

C. Not all arithmetic Fuchsian groups are directly related to triangle ones, though Jacquet-Langland correspondence suggests some relationship at least on the level of representations and underlying algebraic varieties in the $SL_2$ case. In any case, to class C belong those $\alpha$'s and continued fraction expansions for which the corresponding differential equation has an arithmetic monodromy group. Multidimensional arithmetic groups, particularly Picard groups and associated Pochhamer differential equations provide *classes* of continued (more precisely, multidimensional continued) fractions corresponding to periods on algebraic surfaces and varieties.

In applications to diophantine approximations, a particular attention is devoted to three-term linear recurrences like:

$$n^d \cdot u_n = P_d(n) \cdot u_{n-1} - Q_d(n) \cdot u_{n-2} : n \geq 2$$

for $d \geq 2$. Apart from trivial cases (reducible to generalized hypergeometric functions), our conjectures claim that for every $d \geq 1$, there are only finitely many classes of such recurrences and that they all correspond to deformations of algebraic varieties.

For $d = 2$ (second order equations) we have classified nontrivial three-term recurrences whose solutions are always nearly integral, assuming our integrality conjectures. Most of these recurrences are useless in arithmetic applications. There are a few new ones that give some nontrivial results. Among these recurrences are the following:

i) $2n^2u_n = 2(-15n^2+20n-7)\cdot u_{n-1}+(3n-4)^2\cdot u_{n-2}$;
ii) $3n^2u_n = (-12n^2+18n-7)\cdot u_{n-1}+(2n-3)^2\cdot u_{n-2}$;
iii) $n^2u_n = (-12n^2+18n-7)\cdot u_{n-1}+(2n-3)^2\cdot u_{n-2}$;
iv) $n^2\cdot u_n = (56n^2-70n+23)\cdot u_{n-1}-(4n-5)^2\cdot u_{n-2}$.

There is a larger class of rank $r>2$ linear recurrences of the form

$$n^2\cdot u_n = \sum_{k=1}^{r} A_k(n)\cdot u_{n-k},$$

all solutions of which are nearly integral. Many of these recurrences (like iii) above) give rise to new irrationalities. E.g. we present the following new globally nilpotent equation ($r=3$):

$$4x(x^3+16x^2+77x-2)y''+8(2x^3+24x^2+77x-1)y'+(9x^2+70x+84)y=0.$$

Recently, studying Lamé equations we discovered new classes of explicit continued fraction expansions of classical special functions related to arithmetic problems above. These continued fractions expansions generalize many Stieltjes-Roger's continued fraction expansions.

Stileltjes-Roger's expansions [50-54] include the examples:

$$\int_0^\infty sn(u,k^2)e^{-uz}du = \frac{1}{z^2+a-}\,\frac{1\cdot 2^2\cdot 3k^2}{z^2+3^2a-}\,\frac{3\cdot 4^2\cdot 5k^2}{z^2+5^2a-}\,\frac{5\cdot 6^2\cdot 7k^2}{z^2+7a^2-\ldots}$$

$$z\int_0^\infty sn^2(u,k^2)e^{-uz}du = \frac{2}{z^2+2^2a-}\,\frac{2\cdot 3^2\cdot 4k^2}{z^2+4^2a-}\,\frac{4\cdot 5^2\cdot 6k^2}{z^2+6^2a-}\,\frac{6\cdot 7^2\cdot 8k^2}{z^2+8^2a-\ldots} \tag{4.1}$$

$a=k^2+1$.

In the case of expansion (4.1) the approximations $P_m/Q_m$ to the integral in the left hand side of (4.1) are determined from a three-term linear recurrence satisfied by $P_m$ and $Q_m$

$$(2m+1)(2m+2)\phi_{m+1}(z) = (z+(2m+1)^2a)\phi_m(z)-2m(2m+1)k^2\phi_{m-1}(z).$$

Here $\phi_m = P_m$ or $Q_m$, and $Q_m$ are orthogonal polynomials. The generating function of $Q_n$ satisfy a Lamé equation in the algebraic form with a parameter $n=0$. Here $z$ plays a role of the accessory or spectral parameter in the Lamé equation, and the corresponding solutions is

$$y(x) = \sum_{m=0}^{\infty} Q_m(z)\cdot x^m$$

the only solution regular at $x=0$. The generating function of $P_m$ is a regular at $x=0$ solution of nonhomogeneous Lamé equation.

These special continued fraction expansions can be generalized to continued fraction expansions associated with any Lamé equation with an arbitrary parameter $n$.

For $n=0$ these closed form expressions represent the Stieltjes-Rogers expansions. For $n=1$ two classes of continued fractions from [32, §13] have arithmetic applications, because for three values of the accessory parameter $H$ (corresponding to $e_i$-nontrivial

2nd order points) the Lamé equation is a globally nilpotent one and we have $p$-adic as well as archimedean convergence of continued fraction expansions. This way we obtain the irrationality and bounds on the measure of irrationality of some values of complete elliptic integrals of the third kind, expressed through traces of the Floquet matrices. Similarly, for an arbitrary integral $n \geq 1$, among continued fraction expansions, expressed as integrals of elliptic $\theta$-functions, there are $2n+1$ cases of global nilpotence, when continued fractions have arithmetic sense and orthogonal polynomials have nearly integral coefficients.

Among new explicit continued fraction expansions is the expansion of the following function generalizing Stieltjes-Rogers:

$$\int_0^\infty \frac{\sigma(u-u_0)}{\sigma(u)\sigma(u_0)} e^{\zeta(u_0)u} du,$$

or

$$\int_\omega^{\omega+\omega'} \frac{\sigma(u-u_0)}{\sigma(u)\sigma(u_0)} e^{\zeta(u_0)u} du,$$

as a function of $x = \wp(u_0)$. In Jacobi's notations this function can be presented as

$$\int_0^\infty \frac{H(u+u_0)}{\Theta(u)} e^{-uZ(u_0)} du,$$

where $\Theta$ and $H$ are Jacobi's notations for functions.

The three-term linear recurrence determining the $J$-fraction for the corresponding orthogonal polynomials has the following form:

$$Q_N(x) = Q_{N-1}(x) \cdot \{(l+k^2) \cdot (N-1)^2 + x\}$$
$$+Q_{N-2}(x) \cdot k^4 \cdot (N-1)^2 \cdot N \cdot (N-\frac{1}{2}) \cdot (N-\frac{3}{2}) \cdot (N-\frac{5}{2}).$$

Here $x = sn^2(u_0; k^2)$.

The more general $J$-fraction of the form

$$\ddots b_{n-1} + x - \cfrac{a_{n-1}}{b_n + x - \cfrac{a_n}{b_{n+1} + x - \ddots}},$$

with

$$a_n = k^4 \cdot n(n+1) \cdot (n+\frac{1}{2})(n-\frac{1}{2}) \cdot \{(n-1)\cdot(n-\frac{1}{2}) - \frac{m\cdot(m+1)}{4}\};$$
$$b_n = (1+k^2)\cdot(n-1)^2 : n \geq 2$$

is convergent to the integral of the form

$$\int_0^\infty \prod_{i=1}^m \frac{H(u-u_i)}{\Theta(u)\theta(u_i)} e^{-Z(u_i)u} du.$$

The generating function of the corresponding orthogonal polynomials is expressed in terms of solutions of a Lamé equation with parameter $m \geq 1$.

These continued fraction expansions might be the only new additions to cases, when both the function is explicitly known (as an integral of classical functions) and its continued fraction is known.

# 5 Archimedean and P-adic Quadratic Period Relations á la Ramanujan.

Let us turn to applications of arithmetic differential equations, combined with complex multiplication, to diophantine approximations.

As we had stated above, the only arithmetically interesting linear differential equations are globally nilpotent ones. As we have conjectured earlier, these equations are exactly Picard- Fuchs equations of deformation of period (Hodge) structure of algebraic varieties. The first nontrivial case is that of curves, and in this category, elliptic curves are the most widely studied. Deformation of periods of elliptic curves are described by linear differential equations, uniformized by classical congruence subgroups. The full $\Gamma(1)$ group leads to the modular invariant $J = J(\tau)$, and $\Gamma(2)$ to the invariant $\lambda = k^2(\tau)$.

To be more specific and more general, we remind the primitives from the uniformization theory. If $\Gamma$ is an arithmetic group (an arithmetic Fuchsian subgroup of $SL_2(\mathbf{R})$, and $\phi = \phi(\tau)$ is the corresponding automorphic function of $\Gamma$ on $H$, then the function inverse to $\phi(\tau = \tau(\phi))$, is represented as a ratio of two solutions of a second order linear differential equation

$$(\frac{d^2}{d\pi^2} + R(\pi))y = 0$$

with algebraic function coefficients over $\overline{\mathbf{Q}}(\pi)$.

If the genus of $\Gamma$ is zero, the equation has rational function coefficients.

For triangle groups the corresponding linear differential equations are Gauss hypergeometric equations. For 4 triangle subgroups, commensurable with the full modular group, one arrives at 4 theories of hypergeometric function representation of periods of elliptic curves (corresponding to low level structures on these curves).

These 4 theories of hypergeometric function representations are all related by modular identities of a relatively simple form, such as a well known expression of $J(\tau)$ in terms of $\lambda(\tau)$:

$$J(\tau) = 2^8 \cdot \frac{(1-\lambda+\lambda^2)^3}{\lambda^2 \cdot (1-\lambda)^2}.$$

These and other few (from the whole garden of) modular identities are easily translated into the hypergeometric identities between the corresponding representation of periods.

In fact, they all are consequences of simple fractional transformations and a single quadratic relation valid for a large class of hypergeometric functions:

$${}_2F_1(2a, 2b; c-a-b; z) = {}_2F_1(a, b; c-a-b; 4z(1-z)).$$

This or another way such identities were picked up by Ramanujan, [58], see exposition in [48], who also looked at specializations of hypergeometric representation of periods and quasiperiod is of foliations of elliptic curves to curves with complex multiplication. At curves with complex multiplication, modular functions (and combinations of their derivatives) are known to take algebraic values. These algebraic expressions, plugged into hypergeometric functions, lead to hypergeometric function representations of $\Lambda/\Phi$ (and other numbers connected with logarithms of algebraic numbers) as values of (rapidly) convergent hypergeometric series. All necessary algebraic and complex multiplication statements can be found in Weil's book [59].

To introduce Ramanujan's series we first need Eisenstein's series:

$$E_k(\tau) = 1 - \frac{2k}{B_k} \cdot \sum_{n=1}^{\infty} \sigma_{k-1}(n) \cdot q^n$$

for $\sigma_{k-1}(n) = \sum_{d|n} d^{k-1}$, and $q = e^{2\pi i \tau}$. In the $E_k(\tau)$ notations, the quasiperiod relation is expressed by means of the function

$$s_2(\tau) \stackrel{\text{def}}{=} \frac{E_4(\tau)}{E_6(\tau)} \cdot \left(E_2(\tau) - \frac{3}{\pi Im(\tau)}\right), \tag{5.1}$$

which is nonholomorphic but invariant under the action of $\Gamma(1)$.

Ramanujan proved that this function admits algebraic values whenever $\tau$ is imaginary quadratic. Moreover, Ramanujan transforms these relations into rapidly convergent generalized hypergeometric representation of simple algebraic multiples of $1/\pi$. To do this he used only modular functions and hypergeometric function identities. Let us start with Ramanujan's own favorite [58]:

$$\frac{9801}{2\sqrt{2\pi}} = \sum_{n=0}^{\infty} \{1103 + 26390n\} \frac{(4n)!}{n!^4 \cdot (4 \cdot 99)^{4n}}.$$

The reason for this representation of $1/\pi$ lies in the representation of $(K(k)/\pi)^2$ as a ${}_3F_2$–hypergeometric function. Apparently there are four classes of such representations [48] all of which were determined by Ramanujan: all based on four special cases of Clausen identity of a hypergeometric function (and all represented by Ramanujan):

$$F(a, b; a + b + \frac{1}{2}; z)^2 = {}_3F_2\left({2a, a+b, 2b \atop a+b+\frac{1}{2}, 2a+2b} \Big| z\right).$$

The Clausen identity gives the following ${}_3F_2$–representation for an algebraic multiple of $1/\pi$, following from (5.1):

$$\sum_{n=0}^{\infty} \{\frac{1}{6}(1 - s_2(\tau)) + n\} \cdot \frac{(6n)!}{(3n)! n!^3} \cdot \frac{1}{J(\tau)^n} \tag{5.2}$$

$$= \frac{(-J(\tau))^{1/2}}{\pi} \cdot \frac{1}{(d(1728 - J(\tau))^{1/2}}.$$

Here $\tau = (1 + \sqrt{-d})/2$. If $h(-d) = 1$, then the second factor in the right hand side is a rational number. The largest one class discriminant $-d = -163$ gives the most rapidly convergent series among those series where all numbers in the left side are *rational*:

$$\sum_{n=0}^{\infty} \{c_1 + n\} \cdot \frac{(6n)!}{(3n)! n!^3 (-640,320)^n} = \frac{(640,320)^{3/2}}{163 \cdot 8 \cdot 27 \cdot 7 \cdot 11 \cdot 19 \cdot 127} \cdot \frac{1}{\pi}.$$

Here

$$c_1 = \frac{13,591,409}{163 \cdot 2 \cdot 9 \cdot 7 \cdot 11 \cdot 19 \cdot 127}$$

and $J(\frac{1+\sqrt{-163}}{2}) = -(640,320)^3$.

Ramanujan provides instead of this a variety of other formulas connected mainly with the tree other triangle groups commensurable with $\Gamma(1)$. All four classes of ${}_3F_2$ hypergeometric functions (that are squares of ${}_2F_1$–representations of complete elliptic integrals via the Clausen identity). These are

$${}_3F_2\left({1/2,1/6,5/6 \atop 1\quad ,1}\Big| x\right) = \sum_{n=0}^{\infty} \frac{(6n)!}{(3n)!n!^3}\left(\frac{x}{12^3}\right)^n$$

$${}_3F_2\left({1/4,3/4,1/2 \atop 1\quad ,1}\Big| x\right) = \sum_{n=0}^{\infty} \frac{(4n)!}{n!^4}\left(\frac{x}{4^4}\right)^n$$

$${}_3F_2\left({1/2,1/2,1/2 \atop 1\quad ,1}\Big| x\right) = \sum_{n=0}^{\infty} \frac{(2n)!^3}{n!^6}\left(\frac{x}{2^6}\right)^n$$

$${}_3F_2\left({1/3,2/3,1/2 \atop 1\quad 1}\Big| x\right) = \sum_{n=0}^{\infty} \frac{(3n)!}{n!^3} \cdot \frac{(2n)!}{n!^2}\left(\frac{x}{3^3 \cdot 2^2}\right)^n.$$

Representations similar to (3.2) can be derived for any of these series for any singular moduli $\tau \in \mathbb{Q}(\sqrt{-d})$ and for any class number $h(-d)$, thus extending Ramanujan list [58] ad infinum.

Ramanujan's algebraic approximations to $1/\pi$ can be extended to the analysis of linear forms in logarithms arising from class number problems. All of them are natural consequences of Schwarz theory and the representation of the function inverse to the automorphic one (say $J(\tau)$) as a ratio of two solutions of a hypergeometric equation. One such formula is

$$\pi i \cdot \tau = \ln(k^2) - \ln(16) + \frac{G(\frac{1}{2}, \frac{1}{2}; 1; k^2)}{F(\frac{1}{2}, -\frac{1}{2}; 1; k^2)},$$

and another is Fricke's

$$2\pi i \cdot \tau = \ln(J) + \frac{G(\frac{1}{12}, \frac{5}{12}; 1-; \frac{12^3}{J})}{F(\frac{1}{12}, \frac{5}{12}; 1; \frac{12^3}{J})}.$$

Here $G(a,b;c;x) = \sum_{n=0}^{\infty} \frac{(a)_n(b)_n}{(c)_n n!} \cdot \{\sum_{j=0}^{n-1}(\frac{1}{a+j} + \frac{1}{b+j} - \frac{2}{c+j})\}$ is the hypergeometric function (of the second kind) in the exceptional case, when there are logarithmic terms. Similar theory can be constructed for all arithmetic triangle groups [48]. The basic object here is the automorphic function $\phi(\tau)$ for the arithmetic group $\Gamma$ normalized by its values at vertices.

An analog of $s_2(\tau)$ that is a nonholomorphic automorphic form for $\Gamma$ is

$$\frac{1}{\phi'(\tau)} \cdot \left\{\frac{\phi''(\tau)}{\phi'(\tau)} - \frac{i}{Im(\tau)}\right\}.$$

For $\phi(\tau) = J(\tau)$ one gets $s_2(\tau)$.

For example, let us look at a quaternion triangle group (0;3;22,6,6). In this case, instead of elliptic Schwarz formula one has the following representation of the normalized automorphic function $\phi = \phi(\tau)$ in $H$ in terms of hypergeometric functions:

$$\frac{\tau + i(\sqrt{2}+\sqrt{3})}{\tau - i(\sqrt{2}+\sqrt{3})} = -\frac{3^{1/2}}{2^2 \cdot 2^{1/6}} \cdot \{\frac{\Gamma(1/3)}{\sqrt{\pi}}\}^6 \cdot \frac{F(\frac{1}{12}, \frac{1}{4}; \frac{5}{6}; \phi)}{\phi^{1/6} \cdot f(\frac{1}{4}, \frac{5}{12}; \frac{7}{6}; \phi)}.$$

Thus the role of $\pi$ in Ramanujan's period relations is occupied in $(0,3;2,6,6)$−case by the transcendence $\{\frac{\Gamma(1/3)}{\pi}\}^6$.

In the case $0,3;2,4,6)$−group the representation of $\phi = \phi(\tau)$ is

$$\frac{(\sqrt{3}-1)\tau - i\sqrt{2}}{(\sqrt{3}-1)\tau + i\sqrt{2}} = -2(\sqrt{3}-\sqrt{2})\frac{\Gamma(\frac{1}{24})\Gamma(-\frac{5}{24})}{\Gamma(-\frac{13}{24})\Gamma(-\frac{17}{24})} \cdot \phi^{1/2} \cdot \frac{F(\frac{13}{24}, \frac{17}{24}; \frac{3}{2}; \phi)}{F(\frac{1}{24}, \frac{5}{24}; \frac{1}{2}; \phi)}.$$

This leads to a new transcendence:

$$\frac{\Gamma(\frac{1}{24})^4}{\{\Gamma(\frac{1}{3})\Gamma(\frac{1}{4})\}^2}.$$

Thus, generalizations of Ramanujan identities allows us to express constants, such as $\pi$ and other $\Gamma$−factors, as values of rapidly convergent series with nearly integral coefficients in a variety of ways, with convergence improving as the discriminant of the corresponding singular moduli increases.

Rapidly convergent ${}_2F_1$ and ${}_3F_1$ representations of multiples of $1/\pi$ and other logarithms can be and are used for diophantine approximations to corresponding constants in the manner described for globally nilpotent equations. For this one constructs, starting from hypergeometric functions themselves, hypergeometric representation of Padé approximations to them. This specialization of these approximations to complex multiplication points give nearly integral sequences of numerators and denominators in the dense approximations to corresponding constants. Such dense sequences of approximations are used to determine the measure of irrationality (or to prove irrationality) of classical constants. We were conducting extensive computations in this direction, particularly for $\pi$, $\pi/\sqrt{3}$ and $\pi/\sqrt{2}$, and an interesting phenomenon was discovered. Apparently there is a large cancellation (common factors) between numerators and denominators in the sequences of dense approximations, as defined by the corresponding linear recurrences. Also we found some interesting congruences for these dense approximations that allow us to improve measures of irrationalities obtained using these sequences. These congruences have a definitive analytic $p$-adic sense.

Indeed, in addition to archimedean period relations in the complex multiplication case there are corresponding nonarchimedean ($p$−adic) relations reflecting the same modular numbers. These $p$-adic evaluations indicate the possibility of existence of $p$-adic interpretation of hypergeometric identities. Several attempts to give such interpretation were undertaken. One of the more successful is the Koblitz-Gross formula [73] giving $p$-adic interpretation of Gauss sums for Fermat curves in terms of Morita's $p$-adic $\Gamma$ - and $B$ - functions formulas as $p$−adic analogs of Selberg-Chowla formula for periods of elliptic curves with complex multiplication.

In our applications to congruences satisfied by hypergeometric approximations to multiples of $1/\pi$, we do not need a $p-$adic values of the full series, but rather congruences satisfied by truncated hypergeometric series that can be directly interpreted through Hasse invariants and traces of Frobenius.

We briefly describe the background of congruences, taking as our initial model the Legendre form of elliptic curves (and of their periods).

Elliptic curves in the Legendre form are given by the following cubic equation:

$$y^2 = x \cdot (x-1) \cdot (x-\lambda). \tag{5.3}$$

Legendre notations for periods of this curve (= complete integrals of the first kind) and quasiperiods (= complete integrals of the second kind) are, correspondingly,

$$K(\lambda), K'(\lambda)$$

and

$$E(\lambda), E'(\lambda),$$

where

$$K(\lambda) \stackrel{\text{def}}{=} \frac{\pi}{2} \cdot {}_2F_1(\frac{1}{2}, \frac{1}{2}; 1; \lambda),$$

$$E(\lambda) \stackrel{\text{def}}{=} \frac{\pi}{2} \cdot {}_2F_1(-\frac{1}{2}, \frac{1}{2}; 1; \lambda).$$

Similarly:

$$K'(\lambda) = K(\lambda'),$$

$$E'(\lambda) = E(\lambda')$$

for $\lambda + \lambda' = 1$.

We denote

$$K_\lambda = \frac{d}{d\lambda} K(\lambda),\ K'(\lambda), \text{ etc.}$$

The classical Legendre identity

$$K \cdot E' + K' \cdot E - K \cdot K' \equiv \frac{\pi}{2},$$

is equivalent to a simple Wronskian relation for the hypergeometric equation corresponding to the function

$$F(\lambda) \stackrel{\text{def}}{=} {}_2F_1(\frac{1}{2}, \frac{1}{2}; 1; \lambda) = \sum_{n=0}^{\infty} \binom{2n}{n}^2 (\frac{\lambda}{2^4})^n. \tag{5.4}$$

This Wronskian relation is

$$F \cdot F'_\lambda + F' \cdot F^\lambda = \frac{1}{\lambda \cdot \lambda' \cdot \pi} \tag{5.5}$$

(or $K \cdot K'_\lambda + K' \cdot K_\lambda = \frac{1}{2 \cdot \lambda \cdot \lambda'} \cdot \frac{\pi}{2}$).

Over finite fields, there is a well known relation between Hasse invariants and mod $p$ reduction of solutions of the (Picard - Fuchs = Legendre) period linear differential equation. Such a relationship is very general, and we recommend Clemen's book [70] or original Manin's papers [71-72] where such relations are derived via Serre's duality. For elliptic curve in the Legendre form mod $p$ interpretation is particularly easy to express in terms of Legendre function $F(\lambda)$. If one looks at hypergeometric equation satisfied by $F(\lambda)$:

$$\lambda(1-\lambda)\frac{d^2y}{d\lambda^2}+(1-2\lambda)\frac{dy}{d\lambda}+\frac{1}{4}y=0,$$

then this equation is globally nilpotent mod $p$ (as Picard-Fuchs), but does not have two solutions defined mod $p$. Thus, there is a preferred (unique) solution mod $p$. To obtain this polynomials solution one has to reduce all coefficients of the power series expansion of $F(\lambda)$ mod $p$, and then delete all coefficients that follow two consecutive zeroes. This way one arrives to a polynomials mod $p$, known as Hasse-Deuring polynomial:

$$H_p(\lambda)=\sum_{i=0}^{m}\binom{m}{i}^2\lambda^i,\ m\stackrel{\text{def}}{=}\frac{p-1}{2} \tag{5.6}$$

of degree $m=\frac{p-1}{2}$ in $\lambda$.

This polynomial carries mod $p$ properties of the original elliptic curve (5.1):

Lemma 5.1. The trace $a_p(\lambda)$ of Frobenius of an elliptic curve (5.1) over $\mathbf{F}_p$ for $\lambda\in\mathbf{F}_p$ satisfies the following congruence:

$$a_p(\lambda)\equiv(-1)^m\cdot H_p(\lambda) \bmod p.$$

The number $N_p(\lambda)$ of $\mathbf{F}_p$- rational points on an elliptic curve (5.1) is

$$N_p(\lambda)\equiv 1-(-1)^m\cdot H_p(\lambda) \bmod p.$$

The relationship between $H_p(\lambda)$ (this time a polynomial, not a number) is summarized in the following Tate result.

Lemma 5.2. In the ring of formal power series $\mathbf{F}_p[[\lambda]]$ one has the following decomposition:

$$F(\lambda)=H_p(\lambda)\cdot H_p(\lambda^p)\cdot H_p(\lambda^{p^2})\ldots. \tag{5.7}$$

This identity has, in fact, a full $p$-adic meaning, better represented in the form

$$\frac{F(\lambda)}{F(\lambda^p)}\equiv(-1)^m\cdot a_p(\lambda)$$

closely connected with the problem of canonical lifting of Frobenius and analytic continuation inside the supersingular disks in the $\lambda-$ plane.

The polynomial $H_p(\lambda)$ is a transformation of a Legendre polynomial $P_m(x)$ formally identified as follows:

$$H_p(\lambda)=(\lambda-1)^{-m}\cdot P_m\left(\frac{\lambda+1}{\lambda-1}\right):m=\frac{p-1}{2}.$$

Of course, analogs of Hasse-Dewing polynomials exist for all other models of elliptic curves (e.g. the $J-$ representation of these polynomials was studied by Igusa). All these representations can be transformed one into another by birational correspondences reflected in hypergeometric function identities and congruences.

A variety of congruences on Legendre polynomials (most notably Schur congruences and their generalizations to higher powers of primes, studied by us) are all related to the formal completions of elliptic curves.

A richer variety of congruences occur for values of Legendre polynomials corresponding to specific elliptic curves. These congruences again arise from formal groups of these elliptic curves, but are now directly expressed in terms of traces of Frobenius. For higher and composite radices such congruences are known Aitken-Swinnerton-Dyer congruences.

We pay special attention to curves (5.1) with complex multiplication (i.e. when $\lambda$ is a singular moduli). In this case when the curve $E$ has complex multiplication in the imaginary quadratic field $K$, the trace of Frobenius, or the value $H_p(\lambda$ of Hasse-Deuring polynomial has a variety of arithmetic interpretations. It is easier to look at one-class fields $K$. The half of the primes $p$ are supersingular for the elliptic curve (5.1), i.e.

$$H_p(\lambda) \equiv 0 \bmod p.$$

These are the primes $p$ that stay prime in $K$. For other good primes $p$, split in $K$, the trace of Frobenius or $H_p(\lambda)$ is explicitly determined from the representation $4p = a^2 + Db^2$, for discriminant $D$ of $K$.

In fact, a variety of algorithms (starting from Jacobi and investigated by Eisenstein and others) use, as a solution to the problem of representing a prime as a binary quadratic form (typically a sum of two squares for $K$ = Gaussian field) expressions for $H_p(\lambda)$.

As a mod $p$ counterpart to $\Gamma$-function representation of periods of elliptic curves with complex multiplication, one can mention similar binomial function (Morita's p-adic B-function) representation of values of $H_p(\lambda)$ at complex multiplication points due to Koblitz-Gross.

We look now at the simplest case of $\lambda = 1/2$. In this case we have

$$H_p(\frac{1}{2}) \equiv 2^m \cdot (-1)^k \binom{2k}{k} \bmod p \tag{5.8}$$

for $m = \frac{p-1}{2}$ and $k = \frac{p-1}{4}$.

The expression (5.8) holds in nonsupersingular case $p = 4k+1$. In the supersingular case $p \equiv 3 (\mathrm{mod} 4)$,

$$H_p(\tfrac{1}{2}) \equiv 0 \bmod p. \tag{5.8'}$$

To have a full mod $p$ analog of Legendre differential equations we also need an interpretation of

$$\frac{d}{d\lambda} H_p(\lambda)$$

at singular moduli $\lambda$. In the case $\lambda = 1/2$ simple Legendre polynomials identities show that

$$\frac{d}{d\lambda} H_p(\frac{1}{2}) \equiv 0 \bmod p \text{ for } p \equiv 1 \bmod 4, \tag{5.9}$$

and

$$\tfrac{d}{d\lambda}H_p(\tfrac{1}{2}) \equiv -2^{-m+1}\cdot(-1)^k\tbinom{2k}{k} \bmod p \tag{5.9'}$$

for $p = 4k+3$, $m = 2k+1$.

Comparing (5.8-8′), and (5.9-9′), we end up with the congruence

$$H_p(\lambda)\cdot\frac{d}{d\lambda}H_p(\lambda)|_{\lambda=1/2} \equiv 0 \bmod p \tag{5.10}$$

for all $p$ ($> 2$).

This congruence is an immediate nonarchimedean counterpart of one of the three (reducible to a single one) original Legendre identities concerning complete integrals of the first and second kind at singular module. In terms of $F(\lambda)$ it is simply

$$F(\lambda)\cdot\frac{d}{d\lambda}F(\lambda)|_{\lambda=1/2} = \frac{2}{\pi}. \tag{5.11}$$

The congruences (5.8) represent congruences on truncated sums of hypergeometric series representing multiples of $1/\pi$ in this particular and all other Ramanujan-like identities.

Before we present these identities, we have to point to the appearance of a new number $\frac{d}{dx}H_p(\lambda)$ at a singular module $\lambda$. Unlike $H_p(\lambda)$, its derivative evaluated at singular moduli lacks immediate arithmetic interpretation. This invariant is associated not with the formal group of an elliptic curve itself, but with the (two dimensional) formal group of an extension of an elliptic curve by an additive group.

That object is parametrized by

$$(\wp(u), \wp'(u), \zeta(u) + z).$$

In the complex multiplication case, $\frac{d}{d\lambda}H_p(\lambda)$ is quite different in supersingular and nonsupersingular cases. In the nonsupersingular case, $\frac{d}{d\lambda}H_p(\lambda)$ can be expressed in terms of trace of Frobenius. In the supersingular case, however, an interpretation of $\frac{d}{d\lambda}H_p(\lambda)$ is more involved and requires a look at $p-$adic $L-$ functions of elliptic curves at (negative) integral points.

With each of the 4 theories of hypergeometric series representations of period relations we associate congruences for values of truncated series. Congruences differ depending on the order of truncation in an obvious sense, i.e. if a few consecutive coefficients in series are zero mod $M$, all higher coefficients are ignored mod $M$. This way one builds a "$p-$adic" interpretation of Ramanujan identities, without changing left hand side (though the full series are meaningless $p-$adically).

We start with the representative theory corresponding to the absolute invariant $J(\tau)$: The "Ramanujan's identities" were

$$\sum_{n=0}^{\infty}\{c_1+n\}\frac{(6n)!}{(3n)!n!^3}\frac{1}{J^n} = \frac{\delta_1}{\pi},$$

where

$$c_1 = \frac{1}{6}(1 - s_2(\tau)),$$

$$\delta_1 = \frac{1}{2}\sqrt{\frac{-J}{d(12^3 - J)}}$$

for $\tau = (1 + \sqrt{-d})/2$, $J = J(\tau)$.

Now truncations of the ${}_3F_2-$ series in $\frac{1}{J}$ can be appropriately determined mod $p$. We put:

$$S_N^{(1)} \stackrel{\text{def}}{=} \sum_{n=0}^{N} \{c_1 + n\} \cdot \frac{(6n)!}{(3n)! \cdot n!^3} \cdot \frac{1}{J^n}.$$

Theorem 5.3. For all good primes $p$,

$$S_N^{(1)} \equiv 0 \bmod p$$

for $[p/6] \leq N < p$.

Let us look at other theories corresponding to congruence subgroups. To simplify notations we denote *all* theories as follows:

$$\sum_{n=0}^{\infty} \{c_1 + n\} \frac{(6n)!}{(3n)! n!^3} \cdot \frac{1}{J^n} = \frac{\delta_1}{\pi};$$

$$\sum_{n=0}^{\infty} \{c_2 + n\} \frac{(4n)!}{n!^4} \cdot \frac{1}{Y^n} = \frac{\delta_2}{\pi};$$

$$\sum_{n=0}^{\infty} \{c_3 + n\} \frac{(2n)!^3}{n!^6} \cdot \frac{1}{X^n} = \frac{\delta_3}{\pi};$$

$$\sum_{n=0}^{\infty} \{c^4 + n\} \cdot \frac{(3n)! \cdot (2n)!}{n!^5} \cdot \frac{1}{Z^n} = \frac{\delta_4}{\pi}.$$

In all these formulas $J$, $Y$, $X$, $Z$ are singular moduli (complex multiplication) and constants $c_1$, $\delta_1$, $c_2$, $\delta_2$, $c_3$, $\delta_3$, $c_4$, $\delta_4$ lie in the corresponding Abelian extensions of field $K$ of complex multiplication. They all are easily explicitly expressed in terms of Eisenstein's nonholomorphic series $s_2(\tau)$. More interestingly the choices of $c_i$, $\delta_i$ are *unique* anyway! This follows from our results on algebraic independence of periods and quasiperiods of elliptic curves with complex multiplication [68], [69].

[This last remark together with congruences and trivial bounds on degrees of $c_i$ and $\delta_i$ gives another purely modular approach to determine $c_i$ and $\delta_i$ as well as other values of nonholomorphic modular functions. In practice, such approach is quite efficient.]

In four theories above the singular moduli have the following expressions:

$$J = J(\tau);$$

$$Y = 2^6 \cdot (2\lambda/\lambda' + 1 + \lambda'/2\lambda)$$

for $\lambda = k^2(\tau)$, $\lambda + \lambda' = 1$;

$$X = 2^4/\lambda \cdot \lambda';$$

$$Z = 3^3/hh',$$

for $J = 3^3 \cdot (9-8h)^3/h^3h'$, $h + h' = 1$.

For each of the theories we have congruences on truncated series:

Theorem 5.4. For any good prime $p$, and the corresponding singular moduli defined mod $p$, the truncated hypergeometric series satisfy the following congruences:

$$S^{(3)}{}_N = \sum_{n=0}^{N} \{c_3 + n\} \cdot \frac{(2n)!^3}{n!^6} \cdot \frac{1}{X^n} \equiv 0 \bmod p^k,$$

for

$$[p^k/2] \leq N \bmod p^k < p^k$$

(i.e. replace $1/\pi$ by 0 mod $p^k$ in the corresponding identity).

For two other theories we have, correspondingly:

$$S_N^{(2)} = \sum_{n=0}^{N} \{c_2 + n\} \cdot \frac{(4n)!}{n!^4} \cdot \frac{1}{Y^n} \equiv 0 \bmod p$$

for

$$[p/4] \leq N \bmod p < p;$$

$$S_N^{(4)} = \sum_{n=0}^{N} \{c_4 + n\} \cdot \frac{(3n)! \cdot (2n)!}{n!^5} \cdot \frac{1}{Z^n} \equiv 0 \bmod p$$

for

$$[p/3] \leq N \bmod p < p.$$

Similar congruences for these two theories, as well as for the previous one, hold with $p$ replaced with $p^k$.

All these congruences are generalized to higher radix congruences by placing truncations at appropriate places. Moreover, there are additional congruences in the supersingular cases. For supersingular primes the truncated series of corresponding ${}_3F_2$ functions vanish mod $p^2$.

For example, let us look at one of Ramanujan's original series representing "pure rational approximation" to $1/\pi$:

$$\sum_{n=0}^{\infty} \binom{2n}{n}^3 \cdot \frac{42n+5}{2^{12n+4}} = \frac{1}{\pi}.$$

Let us look then at a truncated series:

$$\mathcal{F}_N(\lambda) \stackrel{\text{def}}{=} \sum_{n=0}^{N} \{\frac{(\frac{1}{2})_n}{n!}\}^3 \lambda^n.$$

Thus the identity above is

$$[42 \cdot \lambda \cdot \mathcal{F}'_\infty(\lambda) + 5 \cdot \mathcal{F}_\infty(\lambda)]|_{\lambda=1/64} = \frac{16}{\pi},$$

and corresponds to the complex multiplication by $\sqrt{-7}$.

For truncated series we have first the usual congruences:

$$[42 \cdot \lambda \cdot \mathcal{F}'_N(\lambda) + 5 \cdot \mathcal{F}_N(\lambda)]|_{\lambda=1/64} \equiv 0 \bmod p$$

for all

$$\frac{p-1}{2} \le N < p.$$

Next, we have "supersingular congruences":

$$\mathcal{F}_N(\lambda)|_{\lambda=1/64} \equiv 0 \bmod p^2;$$

$$\mathcal{F}'_N(\lambda)|_{\lambda=1/64} \equiv 0 \bmod p$$

for $p \equiv 3, 5, 6 \bmod 7$.

# 6 Elliptic Interpolation Algorithms

We end this presentation with the description of new fast evaluation and interpolation methods on elliptic curves.

To describe our algorithms we look at a general rational function interpolation problem. In the partial fraction representation of rational functions one looks at the rational function with poles only at given (distinct) points: $\alpha_1, \ldots, \alpha_n$. The general form of such a rational function is

$$R(z) = \sum_{i=1}^{n} \frac{x_i}{z - \alpha_i}. \tag{6.1}$$

The evaluation problem for this function consists of simultaneous determination of $n$ values of $R(z)$ at $z = \beta_1, \ldots, \beta_n$ (distinct from $\alpha_i$):

$$y_j = R(z)\,|_{z=\beta_j} = \sum_{i=1}^{n} \frac{x_i}{\beta_j - \alpha_i} : j = 1, \ldots, n. \tag{6.2}$$

It is easy to see that the inverse to this transformation is the following explicit one:

$$x_i = -\sum_{j=1}^{n} \{\frac{P_B(\alpha_i)P_A(\beta_j)}{P'_A(\alpha_i)P'_B(\beta_j)}\} \frac{y_j}{\beta_j - \alpha_i} : \tag{6.3}$$

$i = 1, \ldots, n$. Here, $P_A(x), P_B(z)$ are polynomials of degree $n$ having as roots $\{\alpha_i\}$, and $\{\beta_i\}$, respectively.

Into the scheme (6.1-3) fall discretizations of important one- and multi-dimensional integral transform (with a variety of quadrature approximations methods). Among singular integral transformations that can be described by direct and inverse schemes (6.2-3) the most obvious is the Hilbert transform on a circle. Its proper discretization, and computations via FFTs was described in detail by Henrici. A finite Hilbert transform corresponds in the scheme (6.2-3) to $\alpha_i$ being $N$-th roots of 1, and $\beta_j$ being $N$-th roots of -1. Explicit expressions (after proper normalizations of $x_i$ and $y_i$) of finite

Hilbert transform depends on $N$ being odd or even. For $N$ even finite Hilbert transform is the following:

$$\chi(2n+1) = \frac{2}{N} \sum_{2k \bmod N} f(2k) ctg \frac{\pi}{N}((2n+1) - 2k),$$

$$\chi(2n) = \frac{2}{N} \sum_{2k+1 \bmod N} f(2k+1) ctg \frac{\pi}{N}(2n - (2k+1))$$

for a direct transform, and reversing the order,

$$f(2n+1) = \frac{2}{N} \sum_{2k \bmod N} \chi(2k) ctg \frac{\pi}{N}((2n+1) - 2k),$$

$$f(2n) = \frac{2}{N} \sum_{2k+1 \bmod N} \chi(2k+1) ctg \frac{\pi}{N}(2n - (2k+1))$$

for inverse transform.

Henrici [75] described the reduction of Hilbert transform on an interval to the finite Hilbert transform. This reduction allows for computation of all integrals of the form

$$\frac{1}{\pi} \int_{-1}^{1} \frac{X(t)dt}{t-s} \text{ and } \frac{1}{\pi} \int_{-1}^{1} \frac{X(t)dt}{(1-t^2)(t-s)}.$$

The finite versions of these transforms correspond to schemes (6.2-3) with $\alpha_i$ and $\beta_j$ being roots of Chebicheff polynomials of the first and second kind respectively. In these cases the computational cost of the corresponding finite transforms on a set of $n$ points is $O(n \log n)$, because, as in FFT, any roots of unity are used. For general sets, of points $\alpha_i$ and $\beta_j$, the computational cost is $O(n \log^2 n)$, because one has to use general interpolation algorithms that are of much higher computations cost. In practice, these algorithms are not attractive unless for a very large $n$. In recent papers of Gerasoulis et. al, see [76], the problem of evaluation of rational functions is considered, and several $O(n \log^2 n)$ algorithms are proposed. In these papers, the problem of rational evaluation is referred to as "Trummer's problem" stated by Golub.

We present now several classes of fast rational evaluation algorithms with computational cost $O(n \log n)$, that generalize finite Hilbert transforms. These new transformations correspond to a variety of singular integrals taken over one- dimensional complex continuum and to singular integrals over fractal sets, and to singular integrals with elliptic function kernels. The latter transformations have interesting number theoretic and modular interpretation and we refer to them as Fast Elliptic Number Theoretic Transform (FENTT).

To describe transformations of this class that can be evaluated in $O(n \log n)$ operations, we look at $\alpha_i$ and $\beta_j$ given as roots of polynomials $P_{\mathcal{A}}(z)$ and $P_{\mathcal{B}}(z)$, respectively, where polynomials $P_{\mathcal{A}}$ and $P_{\mathcal{B}}$ correspond to iteration of (fixed) polynomials and rational functions. Thus we start with a sequence of degrees (radices) $D_1, \ldots, D_m$ and with rational functions (polynomials) $R_1(z), \ldots, R_m(z)$ of degrees $D_1$

$,\ldots, D_m$, respectively. The polynomials $P_A$ and $P_B$ have degrees $n = D_1 \ldots D_m$ and have roots as preimages of two distinct points $\alpha$ and $\beta$ under iterated mappings

$$z_{i+1} = R_i(z_i).$$

Thus $P_A(z)$ is defined as (the numerator of) $R_m(R_{m-1}(\ldots(R_1(z))\ldots)) = \alpha$, and $P_B(z)$ as (the numerator of) $R_m(R_{m-1}(\ldots(R_1(z))\ldots)) = \beta$.

The most interesting case is that of $D_1 = \ldots = D_m = D$ and $R_1 = \ldots R_m = R$ being a fixed rational function of degree $D$. The fast algorithm is similar in the flow diagram to the mixed radix FFT schemes. This is a standard divide and conquer scheme in which one replaces the computation of (6.2) with

$$\begin{aligned} P_A(z) &\sim R_m(R_{m-1}(\ldots)) = \alpha \\ P_B(z) &\sim R_m(R_{m-1}(\ldots)) = \beta \end{aligned} \tag{6.4}$$

to the computation of $D$ transforms of the form (6.2) for $D$ pairs:

$$\begin{aligned} P'_A(z) &\sim R_{m-1}(\ldots) = \alpha' \\ P'_B(z) &\sim R_{m-1}(\ldots) = \beta' \end{aligned} \tag{6.5}$$

The mixed radix FFT algorithm is a special case of (6.4-5) with $R_i(z) = z^{D_i}$ : $i = 1, \ldots, m$. The total computational cost of such mixed degree algorithm depends on the computational cost of multiplication of polynomials of degrees $O(D_i)$. In case of $D_i = O(D)$, and of large $m$, the total computational cost of evaluation of (6.2) is $O(mD_1 \ldots D_m)$.

Among rational transforms that fit into this scheme we can mention: a) Hilbert-like transforms for various Julia sets corresponding to polynomial or rational mappings $z \to R(z)$; b) singular integral transformations with elliptic function kernels. In the case a), contours of integration can have arbitrary fractional dimension. In the case b), a continuous analog of the transformation (6.2) is the following:

$$Y(s) = \int_0^{\omega} X(t)\wp(t-s)dt$$

for the Weierstrass elliptic function $\wp(u)$. The latter transform and its discrete versions are particularly well-suited for arithmetic interpretations. If an elliptic curve $\mathbf{E}$ is defined over a finite field $k = \mathbf{F}_p$ (e.g. is a reduction mod $p$ of an elliptic curve over $\mathbf{Q}$), then the set of its $k$-rational points is an Abelian group of order $N_p = p - a_p + 1$ for $|a_p| \leq 2\sqrt{p}$. Moreover, for any integral $a$, $|a| < 2\sqrt{p}$, there is an elliptic curve over $\mathbf{F}_p$ with $N_p = p - a + 1$. Whenever $2^n$ divides $N_p$, one has points of order $2^n$ on an elliptic curve, all defined over $\mathbf{F}_p$. Consequently, the fast evaluation algorithm (6.4-5) can be applied in this case with $D_1 = \ldots = D_m = 4$. The rational function $R(z)$ in this case is the duplication formula for $x$-coordinate in the Weierstrass cubic form of an elliptic curve $E : y^2 = 4x^3 - g_2x - g_3$:

$$R(x) = -2x + \frac{(6x^2 - g_2/2)^2}{4y^2}$$

In (6.4) the choice of $\alpha$ and $\beta$ should be of $x$- coordinates of second order points on $\mathbf{E}$. Whenever prime $p$ is such that $p$ is within distance $2\sqrt{p}$ from a power of 2 (or from a highly composite number) one has a very fast algorithm of rational evaluation and transformation of length $O(p)$ mod $p$.

The same method can be used for a composite number of $M$ if one chooses an appropriate elliptic curve over $\mathbf{Z}/M\mathbf{Z}$, whose reduction mod $p$ for prime factors $p$ of $M$ have highly composite Abelian group of $\mathbf{F}_p$-rational points. Using the standard facts of the distribution of highly composite numbers, see [74], we conclude that with any number $M$ we have FENTT of length $O(M)$ over $\mathbf{Z}/M\mathbf{Z}$ with computational cost $O(M \log M)$. FENTT algorithms are particularly attractive in parallel implementation because they can be executed in parallel for many primes and many elliptic curves with a full result brought together via Chinese remainder theorem.

As an extra application of FENTT, one can use torsion divisors on elliptic curves to decrease the additive complexity of polynomial multiplication and convolution using FENTT.

## REFERENCES

[1] P. Griffiths, Periods of integrals on algebraic manifolds: summary of main results and discussion of open problems, Bull. Amer. Math. Soc., 75 (1970), 228-296.

[2] N. Katz, Nilpotent connections and the monodromy theorem: applications of a result of Turrittin, Publ. Math. I.H.E.S., 32 (1970) 232-355.

[3] C.L. Siegel, Über einige Anwendungen diophantischer Approximationen, Abh. Preuss. Akad. Wiss. Phys. Math. Kl., 1, 1929.

[4] C.L. Siegel, Transcendental Numbers, Princeton University Press, Princeton, 1949.

[5] A.B. Shidlovsky, The arithmetic properties of the values of analytic functions, Trudy Math. Inst. Steklov, 132 (1973), 169- 202.

[6] A. Baker, Transcendental Number Theory, Cambridge University Press, Cambridge, 1975.

[7] G. V. Chudnovsky, On some applications of diophantine approximations, Proc. Nat'l. Acad. Sci. USA, 81 (1984), 1926- 1930.

[8] W. M. Schmidt, Diophantine Approximations, Lecture Notes Math., v. 785, Springer, N.Y. 1980.

[9] A.L. Galočhkin, Lower bounds of polynomials in the values of a certain class of analytic functions, Mat. Sb., 95 (1974), 396-417.

[10] G. V. Chudnovsky, Padé approximations and the Riemann monodromy problem, in Bifurcation Phenomena in Mathematical Physics and Related topics, D. Reidel, Boston, 1980, 448-510.

[11] G. V Chudnovsky, Measures of irrationality, transcendence and algebraic independence. Recent progress, in Journees Arithmetiques 1980 (Ed. by J.V. Armitage), Cambridge University Press, 1982, 11-82.

[12] E. Bombieri, On G-functions, in Recent Progress in Analytic Number Theory (Ed. by H. Halberstram and C. Hooly), Academic Press, N.Y., v. 2, 1981, 1-67.

[13] K. Väänänen, On linear forms of certain class of G-functions and p-adic G-functions, Acta Arith., 36 (1980), 273-295.

[14] G. V. Chudnovsky, On applications of diophantine approximations, Proc. Nat'l. Acad. Sci. USA, 81 (1984), 7261- 7265.

[15] D.V. Chudnovsky, G.V. Chudnovsky, Applications of Padé approximations to diophantine inequalities in values of G- functions, Lecture Notes Math., v. 1135, Springer, N.Y., 1985, 9- 51.

[16] N. Katz, Algebraic solutions of differential equations, Invent. Math., 18 (1972), 1-118.

[17] T. Honda, Algebraic differential equations, Symposia Mathematica, v. 24, Academic Press, N.Y., 1981, 169-204.

[18] B. Dwork, Arithmetic theory of differential equations, ibid., 225-243.

[19] D. V. Chudnovsky, G.V. Chudnovsky, Applications of Padé approximations to the Grothendieck conjecture on linear differential equations, Lecture Notes Math., v. 1135, Springer, N.Y. 1985, 52-100.

[20] D.V. Chudnovsky, G.V. Chudnovsky, Padé approximations and diophantine geometry, Proc. Nat'l. Acad. Sci. USA, 82 (1985), 2212-2216.

[21] J.-P. Serre, Quelques applications du théoreme de densité de Chebtarev, IHES Pulb. Math., 54 (1981), 323-401.

[22] G. Faltings, Eudichkeitssätze für abelsche varietäten über zahlkörpern, Invent. Math., 73 (1983), 349-366.

[23] T. Honda, On the theory of commutative formal groups, J. Math. Soc. Japan, 22 (1970), 213-246.

[24] D.V. Chudnovsky, G.V. Chudnovsky, p-adic properties of linear differential equations and Abelian integrals, IBM Research Report RC 10645, 7/26/84.

[25] D.V. Chudnovsky, G.V. Chudnovsky, The Grothendieck conjecture and Padé approximations, Proc. Japan Acad., 61A (1985), 87-90.

[26] N. Katz, A conjecture in the arithmetic theory of differential equations, Bull. Soc. Math. France, 110 (1982), 203-239; corr., 347-348.

[27] D. V. Chudnovsky, G.V. Chudnovsky, A random walk in higher arithmetic; Adv. Appl. Math., 7 (1986), 101-122.

[28] G.V. Chudnovsky, A new method for the investigation of arithmetic properties of analytic functions, Ann. Math., 109 (1979), 353-377.

[29] C. Matthews, Some arithmetic problems on automorphisms of algebraic varieties, in Number Theory Related to Fermat's Last Theorem, Birkhauser, 1982, 309-320.

[30] B. Dwork, P. Robba, Effective p-adic bounds for solutions of homogeneous linear differential equations, Trans. Amer. Math. Soc., 259 (1980), 559-577.

[31] E. Whittaker, G. Watson, Modern Analysis, Cambridge, 1927.

[32] D.V. Chudnovsky, G.V. Chudnovsky, Computer assisted number theory with applications, Lecture Notes. Math., v. 1240, Springer, N.Y. 1987, 1-68.

[33] D.V. Chudnovsky, G.V. Chudnovsky, Remark on the nature of the spectrum of Lamé equation. Problem from transcendence theory., Lett Nuovo Cimento, 29 (1980), 545-550.

[34] H. Poincaré, Sur les groupes les équations linéaires, Acta Math., 5 (1884), 240-278.

[35] R. Fricke, F. Klein, Vorlesungen über die theorie der Automorphen Functionen, bd. 1., Teubner, 1925.

[36] V. I. Smirunov, Sur les équations differentielles linéaires du second ordre et la théorie des fouctions automorphes, Bull. Soc. Math. 45 (2) (1921), 93-120, 126-135.

[37] L. Bers, Quasiconformal mappings, with applications to differential equations, function theory and topology, Bull. Amer. Math. Soc., 83 (1977), 1083-1100.

[38] L. Keen, H.E. Rauch, A.T. Vasques, Moduli of unctured tori and the accessory parameter of Lamé's equation, Trans. Amer. Math. Soc., 255 (1979), 201-229.

[39] E. Hilb, Lineare Differentialgleichungen im komplexen Gebiet, Enzyklopädie der Math. Wissenschaften II, Band 6, Teubner, 1917, 471-562.

[40] L. A. Takhtadjan, P.G. Zograf, The Liouville equation action—the generating function for accessory parameter, Funct. Anal., 19 (1975), 67-68.

[41] E. G. C. Poole, Introduction to the Theory of Linear Differential Equations, Oxford, 1936.

[42] A. J. Van der Poorten, A proof that Euler missed... Appery's proof of the irrationality $\varsigma(3)$. Math. Intelligeneer, 1 (1978/79), 195-203.

[43] D.V. Chudnovsky, G.V. Chudnovsky, Padé and rational approximations to systems of functions and their arithmetic applications, Lecture Notes Math., v. 1052, Springer, N.Y., 1984, 37-84.

[44] D.V. Chudnovsky, G.V. Chudnovsky, The use of computer algebra for diophantine and differential equations, in Computer Algebra as a Tool for Research in Mathematics and Physics, Proceedings of the New York Conference 1984, M. Dekker, N.Y.(to appear).

[45] B. Dwork, A deformation theory for the zeta function of a hypersurface. Proc. Intern. Congr. Math. Stockholm, 1962, Djursholm, 1963, 247-259.

[46] C. Maclachlan, G. Rosenberg, Two-generator arithmetic Fuchsian groups, Math. Proc. Cambridge Phil. Soc., 93 (l983), 383-391.

[47] K. Takeuchi, Arithmetic Fuchsian groups with signature $(1;e)$ J. Math. Soc. Japan, 35 (1983), 381-407.

[48] D.V. Chudnovsky, G.V. Chudnovsky, Approximations and complex multiplication according to Ramanujan, in Proceedings of the Ramanujan Centenary Conference, Ed. by R. Berndt, Academic Press, (to appear), 97pp.

[49] R. Askey, Orthogonal polynomials and theta functions, Proc. of 1988 AMS Summer School on Theta - Functions (in print).

[50] T.J. Stieltjes, Sur la réduction en fraction continue d'une série procédant suivant les puissances descendantes d'une variable, Ann. Fac. Sci. Toulouse, 3 (1889), 1-17. ≡ Oeuvres, t. II, Groningen, 1918, 184-200.

[51] T. J. Stieltjes, Recherches sur lest fractions continue, ibid, 8 (1894), 1-22; 9 (1895), 1-47 ≡ Ouevres, t. II, Groningen, 1918, 402-566.

[52] L. J. Rogers, On the representation of certain asymptotic series as convergent continued fractions, Proc. London, Math. Soc., (2), 4 (1907), 72-89.

[53] I.J. Schur, Ueber Potenzreihen die im Invern des Einheitskreises beschränkt sind, J. Reing Angev. Math., 147 (1916), 205-232; 148 (1917), 122-145.

[54] L. Carlitz, Some orthogonal polynomials related to elliptic functions, Duke Math. J., 27 (1960), 443-460.

[55] G.V. Chudnovsky, The inverse scattering problem and its applications to arithmetic, algebra and transcendental numbers, Lecture Notes Physics, v. 120, Springer, N.Y., 1980, 150-198.

[56] G.V. Chudnovsky, Rational and Padé approximations to solutions of linear differential equations and the monodromy theory, in Proceedings of the Les Houches International Colloquium on Complex Analysis and Relativistic Quantum Field Theory, Lecture Notes Physics, c. 126, Springer, N.Y., 1980, 136- 169.

[57] G.V. Chudnovsky, Number theoretical applications of polynomials with rational coefficients defined by extremality conditions, in Arithmetic and Geometry, Progress in Mathematics, v. 35, Birkhauser, Boston, 1983, 67-107.

[58] S. Ramanujan, Modular equations and approximations to $\pi$, Collected Papers, Cambridge, 1927, 23-39.

[59] A. Weil, Elliptic Functions According to Eisenstein and Kronecker, Springer, 1976.

[60] C.L. Siegel, Zum Beneise des Starkschen Satzes, Invent. Math. 5 (1968), 180-191.

[61] G.V. Chudnovsky, Padé approximations to the generalized hypergeometric functions I, J. Math. Pures Appl., Paris, 58 (1979), 445-476.

[62] G. Shimura, Automorphic forms and the periods of Abelian varieties, J. Math Soc. Japan, 31 (1979), 561-579.

[63] P. Deligne, Cycles de Hodge absolus et périodes des intégrales des varietés abéiennes, Bull. Soc. Math. de France, Memoire, No 2, 1980, 23-33.

[64] G. Shimura, Introduction to the Arithmetic Theory of Automorphic Forms, Princeton University Press, 1971.

[65] R. Morris, On the automorphic functions of the group $(0,3;l_1,l_2,l_3)$, Trans. Amer. Math. Soc., 7 (l906), 425-448.

[66] K. Takeuchi, Arithmetic triangle groups J. Math. Soc. Japan, 29 (l977), 91-106.

[67] H.P.F. Swimerton-Dyer, Arithmetic groups, in Discrete Groups and Automorphic Functions, Academic Press, 1977, 377-401.

[68] G.V. Chudnovsky, Algebraic independence of values of exponential and elliptic functions, Proceedings of the International Congress of Mathematicians, Helsinki 1979, Acad. Sci. Fennica, Helsinki, 1980, v.1, 339-350.

[69] G.V. Chudnovsky, Contributions to the Theory of Transcendental Numbers, Mathematical Surveys and Monographs, v. 19, Amer. Math. Soc., Providence, R.I., 1984.

[70] C.H. Clemens, A Scrapbook of Complex Curve Theory, Plenum, 1980.

[71] Y.I. Manin, Algebraic curves over fields with differentiation, Izv. Akd. Nauk. SSSR, Ser. Mat. 22 (1958), 737- 756.

[72] Y.I. Manin, The Hasse-Witt matrix of an algebraic curve, ibid., 25 (1961), 153-172.

[73] B. Gross, N. Koblitz, Gauss sums and the $p$ - adic $\Gamma$- function, Ann. Math., 109 (1979), 569-581.

[74] D.V. Chudnovsky, G.V. Chudnovsky, Sequences of numbers generated by addition in formal groups and new primality and factorization tests, Adv. Meth. 7 (1986), 385-434.

[75] P. Henrici, Applied and Computational Complex Analysis, v. 3, John Wiley, 1986.

[76] A. Gerasoulis, M. Grigoriadis, L. Sun, A fast algorithm for Trummer's problem, SIAM J. Ser. Stat. Comput., 8 (1987), s135- s137.

# ITERATION OF TWO-VALUED MODULAR EQUATIONS

Harvey Cohn*
Department of Mathematics
The City College of New York
New York, N.Y., 10031

## 1. Introduction

In a previous series of papers (see [1] and [2], the author considered the iterative computation of $j(zb^n)$ for $n = 1,2,3,...$ starting from the value of $j(z)$ by algebraic equations derived from the modular equation. The original purpose was to apply ring class field theory to imaginary quadratic fields of discriminant $db^n$ for various fundamental discriminants $d(<0)$. The values of $b$ for which this study was possible were severely limited to those for which a "Klein Theory" of invariants [5] existed, namely $b = 2,3,4,5,$ and $7$. Each case was "special" despite the ubiquitous presence of the polyhedral group concept, bringing to mind a remark of C.L. Siegel which W. Magnus [7] quoted in a closely related context: "<u>The mathematical universe is inhabited not only by important species but also by interesting individuals.</u>"

Nevertheless, even without a polyhedral theory, a larger set of 37 cases due to Fricke [6] exists for which an explicit iterative process is possible. The cases are

(1.1) $b = 2,...,21,23,...27,29,31,32,35,36,39,41,47,49,50,59,71$ .

These values of $b$ are characterized by the following property: Let $H$ be the upper half z-plane and let $\Gamma(=PSL(2,Z))$ be the modular group. Then let $\Gamma^0(b)$ be the subgroup of $\Gamma$ for which $j(z/b)$ is invariant and let $\Gamma^*(b)$ be the extension by the symmetry operation

*) Research supported by NSF Grant DMS-8602077

(1.2) $W: z \to -b/z$

(so $\Gamma^*(b) = \Gamma^0(b) \cup W\Gamma^0(b)$) . Then the values of $b$ in (1.1) are precisely those for which $b>1$ and $H/\Gamma^*(b)$ compactifies to a Riemann surface over $H/\Gamma$ of genus 0 , (see [6]). For each of Fricke's 37 cases we shall now describe a method to deduce

(1.3) $j(z) = j(z/b^n)$

iteratively from $j(z)$ . (It should be noted that the theory of $j(zb^n)$ is wholly equivalent under an isomorphism of the group $\Gamma^0(b)$) .

2. The two-valued modular equations.

The function $j(z/b)$ is algebraic over $j(z)$ , satisfying a symmetric modular equation in $X$

(2.1) $\Phi(X,j(z)) = 0$

whose degree in each variable is

(2.2) $\Psi(b) = b\,\Pi(1+1/p),\ p|b$ .

The conjugates are the set of $\Psi(b)$ values (including $j(bz)$,

(2.3a) $j = j((Az+B)/D)$ ,

where $A,B,D$ are integers satisfying

(2.3b) $\gcd(A,B,D) = 1,\quad AD = b,\ 0\le B<D$ .

(This is why $j(z/b)$ and $j(bz)$ have isomorphic theories.) The modular equations, of course, have large coefficients and are too unwieldy for the few cases where they are written down. A more convenient form of the modular equation is the two-valued form in terms of rational functions

(2.4a) $j(z) - j(z/b) = D_b(t,s)$

(2.4b) $j(z)j(z/b) = N_b(t)$ ,

where $t$ and $s$ are parameters describing the hyperelliptic surface

$H/\Gamma^0(b)$ as a two-valued surface over $H/\Gamma^*(b)$ . Thus

(2.5) $$s^2 = P(t)$$

for a polynomial of degree $2g+2$ , with $g$ the genus of $H/\Gamma^0(b)$ . Note that since $j(Wz) = j(-b/z) = j(z/b)$ , then $j(z)$ and $j(z/b)$ are each defined on $H/\Gamma^0(b)$ , (i.e., in $t$ and $s$ ) while the symmetric (single-valued) functions are defined on $H/\Gamma^*(b)$ (i.e., in $t$ ). Thus $D_b(t,s)/s$ is rational in $t$.

Now $j(z)$ and $j(z/b)$ are roots of the equation (in $j$ )

(2.6) $$j^2 - S_b(t)j + N_b(t) = 0 ,$$

where, we now require the (symmetric) trace function

(2.7a) $$j(z) + j(z/b) = S_b(t)$$

(2.7b) $$S_b(t)^2 - D_b(t,s)^2 = 4N_b(t) .$$

The concepts described here were introduced by Fricke [6], but, no doubt owing to practical computability considerations, not a single computation of the equation (2.6) was completed. It is even more remarkable that in all but six cases (b = 39,41,47.50,59,71), Fricke had all the relations needed to deduce (2.6).

In each case a choice of $s$ and $t$ can be made so that the functions $D_b(t,s)$, $N_b(t)$, $S_b(t)$ are all expressible in terms of monic polynomials with integral coefficients. The reason is a patchwork of theorems which involve theta-functions and the discriminant function. The net effect of these cases (see [4]) is to again recall Siegel's concept of the "interesting individual".

## 3. Divisor structure.

Unlike the integrality of coefficients, the divisor structure is not a case of "individuals", but rather a matter of rules. For the (easy) case of $b$ prime, by choice of $t$ the (two) cusps at $z=0$ and $z=\infty i$ are identified by $W$ with the one pole at $t=\infty$ . Thus for $b$

prime we have the following polynomials in $t$ of indicated degree:

(3.1a) $$\deg D_b(t,s)/s = b-g-1$$

(3.1b) $$\deg N_b(t) = b+1$$

(3.1c) $$\deg S_b(t) = b\ .$$

For more general $b$, in $H/\Gamma^0(b)$, $j(z)$ has the poles

(3.2) $$j(z) \approx P_0 + bP_\infty + \sum_{AD=b} \sum_{i=0}^{\phi((A,D))} (D/(A,D))P_{A,D}^{(i)}, (A > 1, D > 1)\ .$$

Here $P_{A,D}^{(i)}$ is a point corresponding to the conjugate in (2.3a). $j((Az+B)/D)$ and $B$ belongs to one of the $\phi((A,D))$ progressions mod $A$ (as indexed by $i$). Also $P_{b,1}^{(1)}$ and $P_{1,b}^{(1)}$ correspond to $P_0$ and $P_\infty$ respectively. From (3.2) and the behavior of $j(z/b)$ $(= j(-b/z))$, we compute the pole structures for the symmetric functions of $j(z)$ and $j(z/b)$. The new "points" are

(3.3a) $$T_\infty = P_0 + P_\infty$$

(3.3b) $$T_A^{(i)} = P_{A,D}^{(i)} + P_{D,A}^{(i)}, (1 < A \le D)\ .$$

Thus the poles are determined in $H/\Gamma^*(b)$ by

(3.4) $$D_b(t,s)/s \approx (b-g-1)T_\infty + \sum \frac{D}{(A,D)} T_A^{(i)} + \begin{cases} \Sigma T_M^{(i)} & (b=M^2 > 4) \\ T_2 & (b=4) \\ 0 & (\text{other } b), \end{cases}$$

(3.5) $$N_b(t) \approx (b+1)T_\infty + \sum \frac{D+A}{(A,D)} T_A^{(i)} + \begin{cases} 2\Sigma T_M^{(i)} & (b=M^2 > 4) \\ T_2 & (b=4) \\ 0 & (\text{other } b), \end{cases}$$

(3.6) $$S_b(t) \approx bT_\infty + \sum \frac{D}{(A,D)} T_A^{(i)} + \begin{cases} \Sigma T_M^{(i)} & (b=M^2 > 4) \\ 0 & (b=4) \\ 0 & (\text{other } b)\ . \end{cases}$$

Note that the summation over $A$ and $D$ is for $1<A<D$, with $b=AD$, and the summation over $i$ is for $\phi((A,D))/2$ arithmetic progressions.

This makes a special case when $b$ is square. (Indeed $b=4$ is even more special because it satisfies $b=M^2$ while $M\equiv -M$ modulo $b$ ). The coefficients of $T_\infty$ of course match (3.1abc). To illustrate for $b=36$, take (3.5). Here

$$N_{36}(t) \approx 37\ T_\infty + 10\ T_2 + 5(T_3^{(1)} + T_3^{(2)}) + 13\ T_4 + 2\ T_6 .$$

Indeed, in our tabulation (see [4]), we see

$$\begin{aligned} N_{36}(t) = {} & (t^6 - 12\ t^4 + 26\ t^3 - 24\ t^2 + 12\ t - 2)^3 . \\ & (t^{18} + 216\ t^{17} - 2844\ t^{16} + 17574\ t^{15} - 69624\ t^{14} \\ & + 200340\ t^{13} - 446394\ t^{12} + 797616\ t^{11} - 1164096\ t^{10} \\ & + 1399232\ t^9 - 1387152\ t^8 + 1129536\ t^7 - 747972\ t^6 \\ & + 395712\ t^5 - 162288\ t^4 + 48936\ t^3 - 9792\ t^2 + 1008\ t - 8)^3 \\ & /((t-2)^2\ (t-1)^{13}\ t^{10}\ (t^2-t+1)^5) \end{aligned}$$

(Note 37 is the degree at $\infty$ while the other poles appear in the denominator).

4. The iteration of the two valued equation.

We now consider the construction of the sequence

(4.1) $$j_n = j(z/b^n)$$

in the context of a function-element on $H/\Gamma^0(b^n)$ , so other branches of the multivalued function of $j(z)$ include $j(zb^n)$ , by analytic continuation. We refer the iteration to the job of constructing a sequence $t_n$ of parameters as follows:

Let $j_0 = j(z)$ and let $t_0$ be one of $\Psi(b)$ values for which

(4.2a) $$j_0^2 - S_b(t_0) j_0 + N_b(t_0) = 0 .$$

The value $j_1 = j(z/b)$, by (2.7a) is given rationally, e.g.,

(4.2b) $$j_1 = S_b(t_0) - j_0$$

since $j_0$ is the other root of (4.2a) as an equation for $j_1$ .

We must associate with $j_1$ a value $t_1$ such that

$$j_1^2 - S_b(t_1)\, j_1 + N_b(t_1) = 0 \tag{4.2c}$$

but it is very important that $t_1 \neq t_0$. (Otherwise when we try to generate $j_2$ we obtain $j_0$ again, see [3] on the "Klein Paradox").

To complete the iteration from $j_0$ to $j_1$, we substitute (4.2b) for $j_1$ in (4.2c) and use (4.2a) to find

$$-j_1 \frac{S_b(t)-S_b(t_0)}{t-t_0} + \frac{N_b(t)-N_b(t_0)}{t-t_0} = 0 . \tag{4.3}$$

This is an equation in $t$ of degree $\Psi(b)-1$ which $t_1$ must satisfy. Having made a choice (which may change under analytic continuation), we find the transition is complete: $(j_0,t_0)$ leads to $(j_1,t_1)$, and in the same fashion,

$$(j_n,t_n) \rightarrow (j_{n+1},t_{n+1}) . \tag{4.4}$$

Note that an alternate form of (4.2b) might be used, namely

$$j_1 = N_b(t_0)/j_0 . \tag{4.5}$$

Of course we make the assumption that we can avoid multiple roots in $t$ in (4.3) at least to the extent of choosing $t_1$ to be different from $t_0$. In other words, our method of iterating (4.1) is "generic". Therefore, in applications to class field theory, this analysis leaves something to be desired, since we work modulo p for the prime p whose ideal is being factored, and exceptional values of $p$ are unavoidable (see [1] and [2]).

## 5. Illustrations

By considering some special cases we shall show how the facility for iteration is generalized, but at the expense of the elegance associated with Klein's polyhedra.

Let us review the earlier method for $b=2$ (see [6]). Here

$$j(z) = F(x) \tag{5.1a}$$

(5.1b) $$j(z/2) = F(1/x)$$

where $F(x)$ is given by

(5.1c) $$F(x) = 64(x+4)^3/x^2 .$$

This set of equations would lead to the "Klein Paradox [3]" if we were to iterate "badly" and write $j(z/4) = j((z/2)/2) = F(1/(1/x)) = j(z)$ . Of course, the trick is that when we start with $z$ we have 3 $(=\Psi(2))$ values of $x$ determined by (5.1a), and, in the second iteration, (to $j(z/4)$) , we must take a different one than in the first. By an elementary calculation, we find [1]

(5.2a) $$j(z) = F(x)$$

(5.2b) $$j(z/2) = F(x^*)$$

(5.2c) $$x^* = ((5x-4) + (x+4)\sqrt{x+1}\ ) / 8x .$$

Similar formulas exist for $b=3,4,5$, and 7 (see [1] and [2]).

The easy case $b=2$ becomes less elegant now under the present set-up (which applies to all of Fricke's cases). From computations in [4], (with a minor change of variables),

(5.3a) $$S_2(t) = t^2 - 495\,t + 54000$$

(5.3b) $$N_2(t) = t^3$$

(5.3c) $$D_2(t,s) = s(t-255)$$

(5.3d) $$s^2 = (t-144)(t-400) .$$

Thus $j(z)$ and $j(z/2)$ are the two roots of

(5.4) $$j^2 - S_2(t)j + N_2(t) = 0 .$$

Hence the iteration coming from equation (4.3) takes the form

(5.5a) $$t_1^2 + t_1(t_0-j_1) + t_0^2 - t_0 j_1 + 495 j_1 = 0 ,$$

where $j_1$ (from equation (4.5)) is given by

(5.5b) $$j_1 = t_0^3/j_0 .$$

We can easily illustrate how the iteration gets started for the roots of $D_2(t,s)$, (compare [4]).

(5.6a) $\underline{t_0 = 400:}$ $j_0 = j(\sqrt{-2}) = 20^3,\ j_1 = j(\sqrt{-2}/2) = 20^3$,

$t = 3800 \pm 26\sqrt{2}\ ,\ j_2 = N_2(t_1)/j_1 = (190 \pm 130\sqrt{2})^3$,

$j(2\sqrt{-2}) = (190+130\sqrt{2})^3,\ j((\sqrt{-2}+1)/2) = (190-130\sqrt{2})^3$;

(5.6b) $\underline{t_0 = 225:}$ $j_0 = j((1+\sqrt{-7})/2) = -15^3,\ j_1 = -15^3$,

$j_1 = j((1+\sqrt{-7})/4 = j(-4/(1+\sqrt{-7})) = j((-1+\sqrt{-7})/2) = j_0$,

$t_1 = -1800 \pm 2025,\ j_2 = N_2(t_1^+)/j_1 = 255^3 = j(\sqrt{-7})$;

(5.6c) $\underline{t_0 = 144:}$ $j_0 = j(i) = 12^3\ ,\ j_1 = N_2(t_0)/j_0 = 12^3$,

$j_1 = j((1+i)/2) = j(-2/(1+i)) = j(i)$,

$t_1 = 792,\ j_2 = N(t_1)/j_1 = 66^3 = j(i/2) = j(2i)$.

For $b=11$, the process involves the solution of a resolvent of degree 11 from the data (computed in [4]).

$$s^2 = t(t^3 - 20\,t^2 + 56\,t - 44)$$

$$N_{11}(t) = (t^4 + 224\,t^3 - 192\,t^2 - 832\,t + 1024)^3$$

$$\begin{aligned} S_{11}(t) = {} & t^{11} - 66\,t^{10} + 1793\,t^9 - 26048\,t^8 + 221056\,t^7 - 1132670\,t^6 \\ & + 3535840\,t^5 - 6683072\,t^4 + 7418752\,t^3 - 4460544\,t^2 \\ & + 1183744\,t - 65536 \end{aligned}$$

$$\begin{aligned} D_{11}(t,s)/2 = {} & (t-16)\,(t-7)\,(t-4)\,(t-2)\,(t-1)\,\cdot \\ & (t^2 - 14\,t + 4)\,(t^2 - 12\,t + 16) \end{aligned}$$

This case might not be entirely beyond elegant methods, by virtue of Klein's analysis of the resolvent [5] for $b=11$, but for $b>11$, the invariant theory terminates abruptly and the iteration for the 37 cases of Fricke becomes rather austere.

## REFERENCES

1. H. COHN, "Iterated ring class fields and the icosahedron", Math. Ann. 255 (1981) 107-122.

2. H. COHN, "Iterated ring class fields and the 168-tesselation", Math. Ann. 270 (1985) 69-77.

3. H. COHN, "Klein's paradox, the icosahedron, and ring class fields", Number Theory, New York (1985), Springer Lect. Notes, Vol. 1135.

4. H. COHN, "The two-valued modular equation", (submitted).

5. R. FRICKE and F. KLEIN, Vorlesungen uber die Theorie der Elliptischen Modulfunctionen, Leipzig, 1892.

6. R. FRICKE, Lehrbuch der Algebra III (Algebraische Zahlen), Braunschweig, 1928.

7. W. MAGNUS, Noneuclidean Tesselations and their Groups, Academic Press, 1974, p. ix.

# Report on Transcendency in the Theory of Function Fields

David Goss

The numbers $\zeta(i) = \sum_{n=1}^{\infty} n^{-i}$, $i$ a positive integer $> 1$, have intrigued mathematicians for many, many years. Researchers have tried to discover whether they are rational, transcendental, rationally related to $\pi^i$, etc.

The first serious advance was due to Euler, who established the following well-known result. This set the tone for all that followed.

**Theorem 1 (Euler):** Let $i$ be an *even* positive integer. Then

$$\zeta(i)/\pi^i \in \mathbb{Q}.$$

For instance: $\zeta(2) = \pi^2/6$, etc.

This result was generalized by C. L. Siegel ([S1]) to the case of *totally-real* number fields $L$ as follows: Let $O_L$ be the ring of algebraic integers of $L$. Let $\mathcal{A} \subseteq O_L$ be an ideal and let $\mathbb{N}(\mathcal{A})$ be the *positive* generator of the *norm* of $\mathcal{A}$. For $i$ a positive integer $> 1$, we set

$$\zeta_L(i) = \sum_{\mathcal{A} \subset O_L} \mathbb{N}(\mathcal{A})^{-i};$$

one checks readily that the sum converges.

**Theorem 2 (Siegel):** Let $d = [L : \mathbb{Q}]$. Then if $i$ is even,

$$\zeta_L(i)/\pi^{di} \in \overline{\mathbb{Q}};$$

where $\overline{\mathbb{Q}}$ = algebraic closure of $\mathbb{Q} \subseteq \mathbb{C}$.

Thus, as $\pi$ is known to be transcendental, we deduce that $\zeta_L(i)$, $i$ even, is also! However, *no* information is given on $\zeta_L(i)$, $i$ *odd* $> 1$.

Let $s \in \mathbb{C}$ with $\mathrm{Re}\{s\} > 1$. It is simple to see that the series $\zeta_L(s)$ converges to a holomorphic function. Such functions have an analytic continuation to a holomorphic function on $\mathbb{C}$ with but a simple pole at $s = 1$, ([L1]). Moreover, such functions have *functional equations* under $s \mapsto 1 - s$. Theorem 2 then becomes equivalent to the fact that $\zeta_L(-i)$, $i$ a non-negative integer, is rational.

Not long ago, R. Apéry ([Ap1]) established that $\zeta(3) \notin \mathbb{Q}$ by elementary methods.

Having now recalled what is known in number fields, let us turn to the main topic of this report: analogs in function fields. After presenting background material, we will describe the elegant theorem of Jing Yu based on the seminal work of Greg Anderson and Dinesh Thakur. This result gives information, for instance, on the analog of $\zeta(i)$ for $i$ *odd*.

Put $A = \mathbf{F}_r[T]$, $r = p^m$, $k = \mathbf{F}_r(T)$ and $K = k_\infty = \mathbf{F}_r((\frac{1}{T}))$. The field $K$ is the completion of $k$ with respect to the valuation $|?|_\infty$, which measures the order of zero with respect to $\frac{1}{T}$ and where $|\frac{1}{T}|_\infty = r^{-1}$. One has $A \subseteq K$ discretely and $K/A$ is compact. This is completely analogous to $\mathbb{Z} \subseteq \mathbb{R}$ discretely and $\mathbb{R}/\mathbb{Z}$ is compact.

In the function field set-up, one can use the above analogy to define "*zeta-values*". Indeed, it is easy to see that, for $i$ a positive integer, the sums

$$\zeta(i) = \sum_{\substack{n \in A \\ n \text{ monic}}} n^{-i}$$

converge to an element of $K$.

Let $t = (r-1) = \#\mathbf{F}_r[T]^*$. In the following, we use congruences modulo $t$, just as above we used congruences modulo $2 = \#\mathbb{Z}^*$.

The next result, due to L. Carlitz ([G1]) in the 1930's, was rediscovered by the author in the 1970's. In it we let $\overline{K}$ be a fixed algebraic closure of $K$ equipped with the canonical extension of $|?|_\infty$.

**Theorem 3 (Carlitz):** There exists a non-zero element $\bar{\pi}$ in $\overline{K}$ such that $\bar{\pi}^t \in K$, and such that for $i \equiv 0 \ (mod\ t)$

$$\zeta(i)/\bar{\pi}^i \in k.$$

The element $\bar{\pi}$ is the "period" of the Carlitz-module $C$, ([G1]).

Now let $L$ be a finite *abelian* extension of $k$ where the prime $\infty$ of $k$ *splits completely*; we abuse language and call $L$ "totally-real". Let $O_L$ = ring of $A$-integers in $L$. For any ideal $I \subseteq O_L$, we let $\mathbb{N}(I)$ be the *monic* generator of its norm. For $i$ a positive integer, we set

$$\zeta_L(i) = \sum_{I \subseteq O_L} \mathbb{N}(I)^{-i};$$

it is trivial to see that these sums converge. Then I have established the following result ([G2]):

**Theorem 4:** Let $i \equiv 0 \ (mod\ t)$ and set $d = [L : k]$. Then

$$(\zeta_L(i)/\bar{\pi}^{di})^2 \in k.$$

It has been shown ([W1]) by Wade that $\bar{\pi}$ is transcendental over $k$. Thus we conclude that $\zeta_L(i)$, $i \equiv 0 \ (mod\ t)$, is transcendental also. Presumably, Theorem 4 should extend to *arbitrary* totally-real extensions.

There is a space, denoted $S_\infty$, which plays the role of "complex exponents" for $\zeta_L(i)$, ([G3]). Indeed, one can define $\zeta_L(s)$ on a "half-plane" of $S_\infty$ and give it an analytic continuation to the total space. However, as yet there is not a functional equation for $\zeta_L(s)$. Thus Theorem 4 has no implications for negative integers, (but see our last remark!).

Next we turn to examine what is known for $i \not\equiv 0 \ (mod\ t)$; where one *cannot* relate the zeta-value to the period $\bar{\pi}$. Here we shall see that the theory of function fields has gone far beyond the theory of number fields. The main result is given in Theorem 5. It was established by Jing Yu based on work of Greg Anderson and Dinesh Thakur.

**Theorem 5 (Yu):** a) $\varsigma(i)$ is transcendental over $k$ for *all* positive $i$.

b) Let $i \not\equiv 0 \ (mod \ t)$. Then $\varsigma(i)/\overline{\pi}^i$ is also transcendental over $k$.

Before sketching the proof of Theorem 5, I want to discuss a very interesting corollary which was point out by D. Thakur. Let $\Pi(x)$ be the *complex analytic* function defined by

$$\Pi(x) = \Gamma(x+1).$$

As is easy to see, $\Pi$ satisfies the functional equation

$$\Pi(x)\Pi(-x) = \frac{\pi x}{\sin(\pi x)}.$$

Moreover, one computes readily that

$$(\Pi'/\Pi)(0) = -\gamma,$$

where $\gamma$ is Euler's constant. Now for function fields, $\Pi(x)$ has an analog in the function

$$\Pi_0(x) = \prod_{n \ monic} (1 + x/n)^{-1}.$$

Indeed, if we let $e(x)$ be the *exponential function* of $C$, one computes ([G5])

$$\prod_{\varsigma \in A^*} \Pi_0(\varsigma x) = \frac{\overline{\pi} x}{e(\overline{\pi} x)}.$$

Therefore, $(\Pi_0'/\Pi_0)(0)$ should be regarded as $-1$ times the *function field* $\gamma$. But it is trivial to see that

$$(\Pi_0'/\Pi_0)(0) = -\varsigma(1);$$

and we see that $\varsigma(1)$ should be regarded as the function field $\gamma$! Thus, we conclude its transcendence (and relationship to $\overline{\pi}$) from Theorem 5!

**Sketch of Proof of Theorem 5:**

1) First of all, by taking the logarithmic derivative of $e(x)$, one establishes Theorem 3 in the classical manner. Indeed, the appropriate values occur in the expansion about $x = 0$ of $e'(x)/e(x) = 1/e(x)$, (see [G1] or [G3]).

2) G. Anderson, ([An1]) has developed a theory of "motives" for function fields. In this theory, one can take the tensor product of $C$ with itself $m$ times to get $C^{\otimes m}$. One sees that $C^{\otimes m}$ is an $m$-dimensional object in that it gives an $A$-action in an $m$-dimensional space. However, it is *rank one* in the sense that the $a$-division points are isomorphic to $A/(a)$. Associated to $C^{\otimes m}$ is an $m$-dimensional exponential function $e_m(X)$, $X = (x_1, \ldots, x_m)$; so $e_1(x) = e(x)$ in the previous notation. Write $e_m(X)$ as

$$e_m(X) = (e_1^m(X), \ldots, e_m^m(X)).$$

The function $e_m(X)$ has a 1-dimensional period lattice $L = A \cdot (\omega_1, \ldots, \omega_m)$. It is a fundamental fact that $\omega_m = \overline{\pi}^m$. Moreover, the exponential function $e_m(X)$ is *surjective* as a function from $\overline{K}^m$ to $\overline{K}^m$. Let $k^s$ be the separable closure of $k \subseteq \overline{K}$. The following result gives the essential set-up for Yu's work.

**Theorem-Lemma 6 (Anderson-Thakur):** There exists a special point $(\ell_1, \ldots, \ell_m) \in \overline{K}^m$ such that

a) $e_m(\ell_1, \ldots, \ell_m) = (y_1, \ldots, y_m) \in (k^s)^m$.

b) $\ell_m = \varsigma(m)$.

c) If $m \not\equiv 0 \ (mod\ t)$, then $(y_1, \ldots, y_m)$ is *not* a torsion point for $C^{\otimes m}$.

We use the symbol $\log_m(X)$ to denote a choice of logarithm for $e_m(X)$; we let $\log = \log_1$.

As an example of the result of Anderson-Thakur, we have the formula of **Carlitz-Thakur**

$$\varsigma(1) = \sum_{j=0}^{\infty} (-1)^j / L_j, \quad L_j = (T^{r^j} - T) \cdots (T^r - T), \quad L_0 = 1;$$
$$e(\varsigma(1)) = 1.$$

The first part of Theorem 5, follows from

**Theorem-Lemma 7 (Yu):** Let $0 \neq (\ell_1, \ldots, \ell_m) \in \overline{K}^m$. If $e_m(\ell_1, \ldots, \ell_m) \in (\overline{k})^m$, then the *last* coordinate is transcendental over $k$.

**Corollary 8:** The last coordinate uniquely determines the special point.

Indeed, if there were two distinct special points with the same last coordinate, one could just subtract to obtain a contradiction to Theorem-Lemma 7.

The functions $\{e_j^m\}$, *called* $E_q$*-functions*, are analogs of classical $E$-functions. Recall that $E$-functions are entire functions which satisfy linear algebraic differential equations. Then Theorem-Lemma 7 follows from the analog of the result of *Schneider-Lang*. This states that, in the space of $m$-variables, $m+1$ $E_q$-functions which take values in a *finite* algebraic extension on an $m$*-dimensional* lattice, must then be *algebraically dependent.*

For instance, if $\varsigma(m)$ is algebraic, one obtains a contradiction using the functions $\{e_1^m, \ldots, e_m^m, x_m\}$ and a certain $m$-dimensional lattice constructed out of $L$.

3) By our construction of special-points, and Corollary 8, one concludes

$\{\varsigma(m), \overline{\pi}^m\}$ are linearly dependent over $k \Leftrightarrow m \equiv 0(t)$.

In fact, otherwise our special point would be a torsion point!

(Indeed, suppose things were 1-dimensional with special point $x$. Then,

$$a \cdot \log x + b\overline{\pi} = 0 \quad \Rightarrow \quad x = e(-b\overline{\pi}/a),$$

which is torsion. The general argument runs in a similar fashion.)

Therefore, we now have two log's which are linearly independent. Then Yu finishes his result by establishing the analog of *Hilbert's 7th problem*: Suppose $(\alpha_1, \ldots, \alpha_m)$, $(\beta_1, \ldots, \beta_m)$ are in $\overline{K}^m$ such that

a) $\{e_m(\alpha_1, \ldots, \alpha_m), e_m(\beta_1, \ldots, \beta_m)\} \subseteq \overline{k}^m$

b) $\alpha_m$, $b_m$ are linearly independent over $k$.

Then they are linearly independent over $k^s$.

To finish, observe that if

$$u\varsigma(m) + v\overline{\pi}^m = 0 \quad \text{for } \{u, v\} \subseteq \overline{k},$$

then, for some $j \geq 0$,

$$u^{p^j}\varsigma(mp^j) + v^{p^j}\overline{\pi}^{mp^j} = 0, \quad \{u^{p^j}, v^{p^j}\} \subseteq k^s.$$

But this cannot happen by the above argument! So if $\lambda = \varsigma(m)/\overline{\pi}^m$ is algebraic, then

$$-\lambda\overline{\pi}^m + \varsigma(m) = 0$$

which is now a contradiction! This completes the sketch of the proof of Theorem 5.

Finally, as mentioned above, one does not yet have a functional equation for $\varsigma(s)$ in the function field case. However, the values of $\varsigma(s)$ at the negative integers can still be shown to be elements of $A$. Let $v \in Spec(A)$. Then one can further establish ([G3]) that $\{\varsigma(-i), i \in \mathbb{N}\}$ interpolate to a continuous $A_v$-valued function on $\mathbb{Z}_p$. Let $i$ be a non-negative integer. Then one sees that

$$\varsigma_v(i) = \sum_{j=0}^{\infty} \left( \sum_{\substack{n \text{ monic} \\ (n,v)=1 \\ \deg(n)=j}} n^{-i} \right) \in A_v .$$

It is known ([G4]) that $\varsigma_v(i) = 0$ for $i \equiv 0 (mod\ t)$. Recently, Yu extended his techniques to show that $\varsigma_v(i)$ is transcendental over $k$ for $i \not\equiv 0 (mod\ t)$!

## References


[An1] Anderson, G.: "$t$-Motives", Duke Math. J., Vol. 53, No. 2, (June 1986) 457-502.

[Ap1] Apéry, R: "Interpolation de fractions continues and irrationalité de certaines constantes", Mathematics. 37-53, CTHS: Bull. Sec. Sci., III, Bid. Nat., Paris 1981.

[G1] Goss, D.: "von-Staudt for $F_q[T]$", Duke Math. J., Vol. 45, (December 1978), 885-910.

[G2] Goss, D.: "Analogies Between Global Fields", Conference Proceedings Canadian Mathematical Society, Vol. 7 (1987), 83-114.

[G3] Goss, D.: "The Arithmetic of Function Fields 2: The 'Cyclotomic' Theory", Journal of Algebra, Vol. 81, No. 1, (March 1983), 107-149.

[G4] Goss, D.: "The Γ-function in the Arithmetic of Function Fields", Duke Journal, Vol. 56, No. 1, (1988), 163-191.

[G5] Goss, D.: "Fourier Series, Measures, and Divided Power Series in the Theory of Function Fields", (preprint).

[L1] Lang, S.: "Algebraic Number Theory, Addison-Wesley, (1970).

[S1] Siegel, C. L.: "Über die analytische Theorie der quadratischen Formen III", Ann. of Math., 38, 1937, 212-291.

[W1] Wade, L. I.: "Certain quantities transcendential over $GF(p^n, x)$", Duke Math. J., **8**, 701-729 (1941).

**EXPONENTIAL SUMS AND FASTER THAN NYQUIST SIGNALING**

*D. Hajela*

Bellcore
435 South Street
Morristown, New Jersey 07960

## 1. Introduction

In this paper we are concerned with the problem of computing the minimal $L_2$ norm over the interval $(-\delta, \delta)$, $0 < \delta \leq 1/2$, of all non-trivial linear combinations of the functions $\exp(2\pi i n \theta)$, $n = 0, 1, \ldots$; where the coefficients in the linear combination are restricted to $0, \pm 1$. The origin of this problem, which is explained more fully after we state the mathematical formulation, arises from certain basic problems in data communications, concerned with studying the behavior of the minimum $L_2$ distance between signals, when data is sent faster than the so called Nyquist rate over an ideal bandlimited channel.

The mathematical formulation of the problem is as follows: For $0 < \delta \leq 1/2$, let,

$$I(\delta) = \inf_{p \in \pi} \left[ \frac{1}{2\delta} \int_{-\delta}^{\delta} | p(\theta)|^2 d\theta \right]^{1/2}$$

where $\pi = \left\{ \sum_{k=0}^{n} \varepsilon_k e^{2\pi i k \theta} \mid n = 0, 1, \ldots ; \varepsilon_k = 0, \pm 1, \varepsilon_0 = 1 \right\}$. We are interested in the behavior of $I(\delta)$ for $0 < \delta \leq 1/2$. Note that by the orthogonality of the exponentials, $I(1/2) = 1$ and further that $I(\delta)$ tends to 0 as $\delta$ tends to 0, since for the polynomial $p(\theta) = 1 - e^{2\pi i \theta}$,

$$\lim_{\delta \to 0} \frac{1}{2\delta} \int_{-\delta}^{\delta} | p(\theta)|^2 d\theta = 0 .$$

Since the exponentials are no longer orthogonal in $L_2(-\delta, \delta)$ for $0 < \delta < 1/2$, this raises the question [5] as to whether there is a $\delta_0 < 1/2$ such that $I(\delta) = 1$ for $\delta_0 \leq \delta \leq 1/2$. The same question can also be asked regarding $I(\delta, L)$ where the definition of $I(\delta, L)$ is exactly the same as $I(\delta)$ except that now, when defining $\pi$, the condition on $\varepsilon_k$ is that $| \varepsilon_k| \leq L$ (and $\varepsilon_0 = 1$). In view of the Stone-Weierstrass theorem it is somewhat surprising that the following is true:

**Theorem 1:** There is a $\delta_0(L) < 1/2$ such that $I(\delta, L) = 1$ for $\delta_0(L) \leq \delta \leq 1/2$.

Before going further, we state the origin and relevancy of the above problem to data communications. It has been known since the 1920's that Nyquist pulses,

$$g(t) = \frac{\sin(\pi t/T)}{\pi t/T}$$

can be used to send data without intersymbol interference over bandlimited channels. Precisely, this means that one sends signals

$$\sum_{n=n_1}^{n_2} a_n\, g(t-nT)$$

if one wants to send binary data $a_n = \pm 1$ over a channel of bandwidth $1/2T$. The absence of intersymbol interference means that the peak of a pulse $g(t-nT)$ is at the zero-crossings of the other pulses $g(t-mT)\,;\, m \neq n$. The above facts have played a major role in the design and implementation of data transmission over the telephone network.

Now suppose we use pulses,

$$g(t) = A\frac{\sin \pi t/T}{\pi t/T}$$

but send such pulses at intervals $R = 2\delta T$ with $0 < \delta < 1/2$ instead of $R = T$. We assume optimum processing of the received signals. We now encounter intersymbol interference and it is natural to use the minimum $L_2$ distance between received signals as a performance criterion. In this case it is easily seen [5] that the minimum distance can be gauged in terms of $I(\delta)$, in the case of binary data being sent. Thus $\delta = 1/2$ corresponds to the classical Nyquist rate, and the question asked earlier about whether $I(\delta) = 1$ in a neighborhood of $\delta = 1/2$ corresponds to asking whether there is a non-degradation of the minimum $L_2$ distance between received signals for rates of transmission somewhat faster than the Nyquist rate.

Besides the above motivation, it should be mentioned that Forney [1] and refinements by others [2], [6] have shown that the bit error rate may be tightly estimated in terms of the minimum distance for high signal to noise ratios. It seems probable that the techniques of this paper can be used to compute the minimum distance when pulse shapes other than Nyquist pulses are used. Since the bit error rate is a basic parameter in gauging the performance of data communication channels, this would be of interest.

Returning to Theorem 1, we shall restrict ourselves to $I(\delta)$, the proofs being the same when $I(\delta, L)$ (which corresponds to multilevel signaling instead of just binary signaling) is considered. Thus we show,

**Theorem 2:** $I(\delta) = 1$ for $.4975 \le \delta \le .5$.

Actually a far stronger result is true [3]: Let $R(\theta) = \sum_{k=0}^{7} (-1)^k e^{2\pi i k\theta}$ and let $0 < \nu < 1/2$ be defined by $\frac{1}{2\nu} \int_{-\nu}^{\nu} |R(\theta)|^2 d\theta = 1$. Then $\nu = .401\ldots$, $I(\delta) = 1$ for $\nu \le \delta \le 1/2$, and $I(\delta) < 1$ for $\delta < \nu$ because $\frac{1}{2\nu} \int_{-\nu}^{\nu} |R(\theta)|^2 d\theta < 1$ for $\delta < \nu$. We prove Theorem 2 here, instead of this stronger result because Theorem 2 is essential to proving the stronger result and because the proof of the stronger result is considerably more difficult. Moreover, many of the elements in the proof of Theorem 2 are similar to those in the proof of the stronger result. Most importantly, the stronger result is a result peculiar to Nyquist pulses and the proof techniques in [3] do not seem to apply to other pulse shapes, while the techniques here seem to apply.

It is not hard to obtain upper bounds on $I(\delta)$, by numerically considering various polynomials. This was done in [5], where the problem of studying $I(\delta)$ was also proposed. Lower bounds on $I(\delta)$ are considerably harder to get and prior to this paper it was only known that $I(\delta) \ne 0$ for all $0 < \delta < 1/2$ [5].

Lastly we prove an extremal form of Theorem 2, which shows that the only reason $I(\delta) = 1$ for $\delta_0 \le \delta \le 1/2$ and some $\delta_0 < 1/2$, is because we have to consider the trivial polynomial $Q(\theta) = 1$:

**Theorem 3:** Given any $1 \le M < \sqrt{2}$ there is a $\delta_0 < 1/2$ such that for $\delta_0 \le \delta \le 1/2$ and any polynomial $Q(\theta) = \sum_{k=0}^{n} \varepsilon_k e^{2\pi i k\theta}$ with $\varepsilon_0 = 1$, $\varepsilon_k = 0, \pm 1$ and $Q(\theta) \ne 1$,

$$\frac{1}{2\delta} \int_{-\delta}^{\delta} |Q(\theta)|^2 d\theta \ge M .$$

**Our notation is standard other than noted below.** A polynomial in this paper shall mean $\sum_{k=0}^{n} \varepsilon_k e^{2\pi i k\theta}$ where $\varepsilon_k = 0, \pm 1$, $\varepsilon_0 \ne 0$. Also $e^{2\pi i\theta}$ **is denoted** $e(\theta)$ and for a function $g$,

$$\hat{g}(x) = \int_{-\infty}^{\infty} g(t)\, e(-tx)\, dt$$

$$\|g\|_2 = \left[ \int_{-\infty}^{\infty} |g(t)|^2 dt \right]^{1/2} .$$

Some of the results of this paper were presented in the conference proceedings [9].

## 2. Non-Degradation of the Minimum Distance For Rates Faster Than the Nyquist Rate:

In this section we give a proof of the result that there is no degradation in the minimum distance between received signals, when signaling at rates somewhat faster than the Nyquist rate. Theorem 2 follows at once by the definition of $I(\delta)$ from the following stronger result:

**Theorem 4:** Let $Q(\theta) = \sum_{0 \le k \le n} \varepsilon_k e^{2\pi i m_k \theta}$ where $0 = m_0 < m_1 < \cdots$, $m_k$ are natural numbers and $\varepsilon_k = \pm 1$. Excluding the trivial case of $Q(\theta) = \pm 1$,

(a) If the minimal gap between consecutive $m_k$ is at least two then,

$$\frac{1}{2\delta} \int_{-\delta}^{\delta} |Q(\theta)|^2 \, d\theta \ge 1$$

for $.393 \cdots \le \delta \le .5$.

(b) If the minimal gap between consecutive $m_k$ is one and the first place where the gap occurs the coefficients corresponding to the exponentials with consecutive $m_k$ have the same sign, then

$$\frac{1}{2\delta} \int_{-\delta}^{\delta} |Q(\theta)|^2 \, d\theta \ge 1$$

for $.38 \cdots \le \delta \le .5$.

(c) If the minimal gap between consecutive $m_k$ is one and the first place where the gap occurs the coefficients corresponding to the exponentials with consecutive $m_k$ have opposite signs then

$$\frac{1}{2\delta} \int_{-\delta}^{\delta} |Q(\theta)|^2 \, d\theta \ge 1$$

for $.4975 \cdots \le \delta \le .5$.

It is easy to see numerically, as mentioned in the introduction, that [5]

$$\frac{1}{2\delta} \int_{-\delta}^{\delta} |R(\theta)|^2 \, d\theta < 1$$

for $\delta \le .4$ where $R(\theta) = 1 + \sum^{7} (-1)^j e^{2\pi i j \theta}$. Thus (a) and (b) of Theorem 4 give a better than expected answer and it is only (c) which does not give as good an answer. It is also quite surprising that there is a distinction between (b) and (c), which is real in view of the numerical example above.

**Proof (Theorem 4):** First we consider part (a). If $Q(\theta)$ has at least $K$ non-zero terms than as a consequence of a theorem of Ingham [4] it follows that

$$\frac{1}{2\delta}\int_{-\delta}^{\delta}|Q(\theta)|^2 d\theta \geq \frac{K}{3}\left[2-\frac{1}{2\delta}\right] .$$

Thus if $K \geq 5$ and $\delta \geq \frac{5}{14}$ we are done. On the other hand assume that $Q(\theta)$ has at most four non-zero terms. It can be shown (see Theorem 7 of the Appendix), that if a polynomial $P(\theta)$ has exactly $n$ non-zero terms,

$$P(\theta)=\sum_{i=1}^{n}\varepsilon_{k_i}\, e(k_i\,\theta) \ , \ \varepsilon_{k_i}=\pm 1$$

then,

$$\frac{1}{2\delta}\int_{-\delta}^{\delta}|P(\theta)|^2 d\theta \geq B_n(\delta)=n\left[1-(n-1)\frac{\sin 2\pi\delta}{2\pi\delta}\right] .$$

Thus,

$$\frac{1}{2\delta}\int_{-\delta}^{\delta}|Q(\theta)|^2 d\theta \geq \min\left(B_2(\delta), B_3(\delta), B_4(\delta)\right)$$

$$\geq 1$$

for $\delta \geq \cdot 393 \cdots$ which proves part (a). We turn to parts (b) and (c). Let $Q(\theta)=\sum_{k=0}^{n}\varepsilon_k\, e(m_k\,\theta)$ where $0=m_0<m_1<\cdots$ and $\varepsilon_k=\pm 1$. Let $k_o$ be the minimal $k$ such that $m_{k+1}-m_k=1$ and let $P(\theta)=\varepsilon_{k_o}\, e(-m_{k_o}\,\theta)\, Q(\theta)$. Clearly,

$$\frac{1}{2\delta}\int_{-\delta}^{\delta}|P(\theta)|^2 d\theta=\frac{1}{2\delta}\int_{-\delta}^{\delta}|Q(\theta)|^2 d\theta \tag{1}$$

and $P(\theta)$ has the form,

$$P(\theta)=1+\varepsilon\, e(\theta)+\sum_{k\neq 0}\varepsilon_k\, e(n_k\,\theta) \tag{2}$$

where $\varepsilon, \varepsilon_k=\pm 1$, $n_k \geq 2$ for $k \geq 1$, $n_k \leq -2$ for $k \leq -1$, $n_{k+1}-n_k \geq 1$ for $k \geq 1$ and $n_k-n_{k-1} \geq 2$ for

$k \le -1$. This is simply by the minimality of $k_o$, and by (1) we need to only get lower bounds on $P(\theta)$. By assumption in part (b) we have $\varepsilon = 1$ in (2) and in part (c) we have $\varepsilon = -1$ in (2). Part (b) will follow from the following theorem:

**Theorem 5:** Let $\varepsilon = 1$ in (2). Then for $P(\theta)$ as in (2) and for $\delta > \frac{1}{4}$:

$$\frac{1}{2\delta}\int_{-\delta}^{\delta} |P(\theta)|^2 d\theta \ge \frac{1}{\sqrt{2}}\left(\left|1 + \frac{\sin 2\pi\delta}{\pi}\left(\frac{1}{2\delta} - \frac{1}{2\delta - 1}\right)\right| - \sum_{k\ge 2}\frac{|\sin 2\pi\delta k|}{\pi}\left(\frac{1}{2\delta k - 1} - \frac{1}{2\delta k + 1}\right)\right).$$

To see how part (b) follows from this, note that

$$\frac{1}{2\delta}\int_{-\delta}^{\delta} |P(\theta)|^2 d\theta \ge f(\delta)$$

where $f(\delta)$ is the estimate obtained in Theorem 5. Since the series defining $f(\delta)$ converges uniformly, it follows that $f(\delta)$ is continuous. Moreover $f(\frac{1}{2}) = \sqrt{2}$. By the intermediate value theorem there is a $\delta_0 < \frac{1}{2}$ such that $f(\delta) \ge 1$ for $\delta_0 \le \delta \le \frac{1}{2}$. A numerical analysis then shows that $\delta_0 = .38 \cdots$. Part (c) follows by a similar analysis from another theorem (see Theorem 8 of the Appendix) whose statement is similar to that of Theorem 5, but is more complicated. This finishes the proof of Theorem 4. □

Theorem 5 follows from the following basic estimate.

**Theorem 6:** Let $P(\theta) = 1 + \varepsilon e(\theta) + \sum_{k\ge 1} \varepsilon_k e(n_k\theta) + \sum_{k<0} \varepsilon_k e(n_k\theta)$ be a polynomial as in (2). Then, for any complex numbers $a_n$ with $\sum_{-\infty}^{\infty} |a_n|^2 < +\infty$,

$$\frac{1}{2\delta}\int_{-\delta}^{\delta} |P|^2 d\theta \ge \left(\sum_{-\infty}^{\infty} |a_n|^2\right)^{-1/2}\left(\left|a_0 + \frac{\varepsilon \sin 2\pi\delta}{\pi}\sum_{-\infty}^{\infty}\frac{a_n}{2\delta - n}\right| - \sum_{k\ne 0}\frac{|\sin 2\pi\delta n_k|}{\pi}\left|\sum_{-\infty}^{\infty}\frac{a_n}{2\delta n_k - n}\right|\right).$$

Theorem 5 follows at once from Theorem 6:

**Proof (of Theorem 5):** Set $a_0 = 1, a_1 = -1$ and all other $a_n = 0$ in the result of Theorem 6. We obtain:

$$\frac{1}{2\delta}\int_{-\delta}^{\delta} |P(\theta)|^2 d\theta \ge \frac{1}{\sqrt{2}}\left(\left|1 + \frac{\sin 2\pi\delta}{\pi}\left(\frac{1}{2\delta} - \frac{1}{2\delta - 1}\right)\right| - \sum_{k\ne 0}\frac{|\sin 2\pi\delta n_k|}{\pi}\left|\frac{1}{2\delta n_k} - \frac{1}{2\delta n_k - 1}\right|\right).$$

Since $n_{k+1} - n_k \ge 1$ for $k \ge 1$ and $n_k - n_{k-1} \ge 2$ for $k < 0$, it follows that for $\delta > \frac{1}{4}$:

$$\frac{1}{2\delta}\int_{-\delta}^{\delta}|P(\theta)|^2\,d\theta \geq \frac{1}{\sqrt{2}}\left(\left|1+\frac{\sin 2\pi\delta}{\pi}\left(\frac{1}{2\delta}-\frac{1}{2\delta-1}\right)\right| - \sum_{k\geq 2}\frac{|\sin 2\pi\delta k|}{\pi}\left(\frac{1}{2\delta k-1}-\frac{1}{2\delta k}\right)\right.$$

$$\left. -\sum_{k\geq 2}\frac{|\sin 2\pi\delta k|}{\pi}\left(-\frac{1}{2\delta k+1}+\frac{1}{2\delta k}\right)\right)$$

$$= \frac{1}{\sqrt{2}}\left(\left|1+\frac{\sin 2\pi\delta}{\pi}\left(\frac{1}{2\delta}-\frac{1}{2\delta-1}\right)\right| - \sum_{k\geq 2}\frac{|\sin 2\pi\delta k|}{\pi}\left(\frac{1}{2\delta k-1}-\frac{1}{2\delta k+1}\right)\right).$$

Note that series in question converge uniformly in $\delta$ and also absolutely (thus rearrangements are allowed). □

Finally we give the proof of Theorem 6. This theorem can be thought of as the basic estimate from which all other estimates can be made to follow.

**Proof (of Theorem 6):** Let $g \in L_2(-\infty, \infty)$ and suppose that $\hat{g}$ is supported in $[-\frac{1}{2}, \frac{1}{2}]$. Let $\eta = 2\delta$ where $0 < \delta \leq \frac{1}{2}$ and let $g_\eta(x) = g(\eta x)$. Then $\hat{g}_\eta(x) = \frac{1}{\eta}\hat{g}(\frac{x}{\eta})$ so that the support of $\hat{g}_\eta$ is $[-\delta, \delta]$. Moreover a simple computation shows,

$$\|g_\eta\|_2^2 = 1/\eta\,\|g\|_2^2\,.$$

Thus $\hat{g}_\eta \in L_2(-\delta, \delta)$ and so upon applying the Cauchy-Schwartz inequality, Plancharel's Theorem, the triangle inequality and the inversion theorem:

$$\|g\|_2\left(\frac{1}{\eta}\int_{-\delta}^{\delta}|P(\theta)|^2 d\theta\right)^{1/2}$$

$$= \|\hat{g}_\eta\|_2\left(\int_{-\delta}^{\delta}|P(\theta)|^2 d\theta\right)^{1/2}$$

$$\geq \left|\int_{-\delta}^{\delta}\Big(\hat{g}_\eta(\theta) + \varepsilon\,\hat{g}_\eta(\theta)\,e(\theta) + \sum_{k\neq 0}\varepsilon_k\,\hat{g}_\eta(\theta)\,e(n_k\theta)\Big)\,d\theta\right|$$

$$\geq \left|\int_{-\delta}^{\delta}\big(\hat{g}_\eta(\theta) + \varepsilon\,\hat{g}_\eta(\theta)\,e(\theta)\big)\,d\theta\right| - \sum_{k\neq 0}\left|\int_{-\delta}^{\delta}\hat{g}_\eta(\theta)\,e(n_k\theta)\,d\theta\right|$$

$$= |g(0) + \varepsilon g(\eta)| - \sum_{k\neq 0}|g(\eta n_k)|\,.$$

Therefore for any $g \in L_2(-\infty, \infty)$ and $\hat{g}$ supported in $[-\frac{1}{2}, \frac{1}{2}]$,

$$\left(\frac{1}{2\delta}\int_{-\delta}^{\delta}|P(\theta)|^2 d\theta\right)^{1/2} \geq \frac{1}{\|g\|_2}\Big(|g(0) + \varepsilon g(\eta)| - \sum_{k\neq 0}|g(\eta n_k)|\Big)\,.$$

By the Paley-Weiner Theorem [7] such a $g$ may be identified with an entire function $g(z)$ with

$|g(z)| \le A\, e^{\pi|z|}$ for some $A > 0$, and in turn such a $g(z)$ may be written as

$$g(z) = \frac{\sin \pi z}{\pi} \sum_{-\infty}^{\infty} \frac{(-1)^n g(n)}{z-n}$$

by Hardy's Theorem [7]. It follows that for any $(a_n)_{n=-\infty}^{\infty}$ with, $\sum_{-\infty}^{\infty} |a_n|^2 < +\infty$ we may define a $g(z)$ by letting $g(n) = a_n$ and then,

$$\frac{1}{2\delta}\int_{-\delta}^{\delta} |P|^2\, d\theta \ge \Big(\sum_{-\infty}^{\infty}|a_n|^2\Big)^{-1/2} \left[\left| a_0 + \varepsilon \frac{\sin 2\pi\delta}{\pi} \sum_{-\infty}^{\infty} \frac{a_n}{2\delta - n}\right| - \sum_{k\ne 0} \frac{|\sin 2\pi\delta n_k|}{\pi} \left| \sum_{-\infty}^{\infty} \frac{a_n}{2\delta n_k - n}\right|\right];$$

since $\|g\|_2^2 = \sum_{-\infty}^{\infty} |g(n)|^2$. □

Finally we prove the extremal form of Theorem 2, namely Theorem 3:

**Proof (Theorem 3):** With the notation as in the statement of Theorem 3, first assume that the minimal gap between non-zero terms in $Q(\theta)$ is at least two. It should be clear that this case can be handled in exactly the same manner as in the proof of Theorem 4(a). If the minimal gap between non-zero terms in $Q(\theta)$ is exactly one, then we may reduce to a polynomial $P(\theta)$ of the form in (2) exactly as in Theorem 4.

For $P(\theta)$ in (2), with $\varepsilon = 1$, we have the estimate in Theorem 5. Clearly this estimate implies the result since the estimate goes to $\sqrt{2}$ as $\delta$ goes to 1/2. For $P(\theta)$ in (2) with $\varepsilon = -1$, we need an appropriate analog of Theorem 8, which is provided by Theorem 10 of the Appendix (also see Lemma 9 and the paragraph preceding it in the Appendix). □

## Appendix

**Theorem 7:** $\displaystyle \frac{1}{2\delta}\int_{-\delta}^{\delta} \Big|\sum_{i\le n} \varepsilon_{k_i}\, e(k_i\,\theta)\Big|^2 d\theta \ge n\left[1 - (n-1)\frac{\sin 2\pi\delta}{2\pi\delta}\right]$, for any $\varepsilon_{k_i} = \pm 1$.

**Proof:** Fix $1 \le k_1 < \cdots < k_n$. We have,

$$\frac{1}{2\delta}\int_{-\delta}^{\delta} \Big|\sum_{i\le n} \varepsilon_{k_i}\, e(k_i\,\theta)\Big|^2 d\theta = n + \sum_{\substack{i\ne j\\ i,j\le n}} \varepsilon_{k_i}\varepsilon_{k_j} \frac{\sin\lambda(k_i-k_j)}{\lambda(k_i-k_j)}$$

$$\left| \sum_{\substack{i,j \le n \\ i \ne j}} \varepsilon_{k_i} \varepsilon_{k_j} \frac{\sin \lambda(k_i - k_j)}{\lambda(k_i - k_j)} \right|$$

$$\le \sum_{i \le n} \sum_{\substack{j \le n \\ j \ne i}} \frac{|\sin \lambda(k_i - k_j)|}{\lambda |k_i - k_j|}$$

$$\le \frac{n}{\lambda} \max_{i \le n} \sum_{\substack{j \le n \\ j \ne i}} |\sin \lambda| \text{ (using the inequality } |\sin nx| \le |n|\,|\sin x| \text{ for integer } n)$$

$$= \frac{n}{\lambda}(n-1)|\sin \lambda| \text{, which completes the proof. } \square$$

**Theorem 8:** Let $\varepsilon = -1$ in (2). Then for $P(\theta)$ as in (2) and $\delta > \frac{2}{5}$,

$$\frac{1}{2\delta}\int_{-\delta}^{\delta} |P(\theta)|^2 d\theta \ge \frac{\sqrt{3}}{\sqrt{10}}\left( \left| 1 - \frac{\sin 2\pi\delta}{\pi}\left( \frac{1}{2\delta} + \frac{1}{2\delta - 1} - \frac{2}{3}\left[ \frac{1}{2\delta + b_1} + \frac{1}{2\delta + b_2} + \frac{1}{2\delta + b_3} \right] \right) \right| \right.$$

$$- \sum_{k \le -6, k \ge 2} \frac{|\sin 2\pi\delta k|}{\pi} \left| \frac{1}{2\delta k} + \frac{1}{2\delta k - 1} - \frac{2}{3}\left( \frac{1}{2\delta k + b_1} + \frac{1}{2\delta k + b_2} + \frac{1}{2\delta k + b_3} \right) \right|$$

$$\left. - \sum_{k \in A} \frac{|\sin 2\pi\delta k|}{\pi} \left| \frac{1}{2\delta k} + \frac{1}{2\delta k - 1} - \frac{2}{3}\left( \frac{1}{2\delta k + b_1} + \frac{1}{2\delta k + b_2} + \frac{1}{2\delta k + b_3} \right) \right| \right)$$

where,

(a) $b_1 = 1, b_2 = 2, b_3 = 3$ and $A = \{-4, -5\}$ if $|n_{-1}| \ge 4$.

(b) $b_1 = 1, b_2 = 2, b_3 = 4$ and $A = \{-3, -5\}$ if $n_{-1} = -3$.

(c) $b_1 = 1, b_2 = 3, b_3 = 4$ and $A = \{-2, -5\}$ if $n_{-1} = -2, |n_{-2}| \ge 5$.

(d) $b_1 = 1, b_2 = 3, b_3 = 5$ and $A = \{-2, -4\}$ if $n_{-1} = -2, n_{-2} = -4$.

**Proof:** First note that all the series in question converge uniformly. This is because:

$$\frac{1}{x} + \frac{1}{x-1} - \frac{2}{3}\left[ \frac{1}{x + b_1} + \frac{1}{x + b_2} + \frac{1}{x + b_3} \right] = \frac{P(x)}{Q(x)}$$

where $Q(x)$ is a polynomial of degree 5 and $P(x)$ is a polynomial of degree 3 and so,

$$\sum_{k \le -6, k \ge 2} \frac{|P(2k\delta)|}{|Q(2k\delta)|}$$

is finite. Also note since $\delta > \frac{2}{5}$ none of the denominators, $2\delta k$, $2\delta k - 1$, $2\delta k + b_1$, $2\delta k + b_2$, $2\delta k + b_3$ vanish. We now show the estimate (a). The others are obtained in an analogous manner. Set $\varepsilon = -1$,

$d=1, a_0=1, a_1=1, a_{-1}=-\frac{2}{3}, a_{-2}=-\frac{2}{3}\ a_{-3}=-\frac{2}{3}$ and all other $a_n=0$ in the estimate for Theorem 6.

We get:

$$\frac{1}{2\delta}\int_{-\delta}^{\delta}|P|^2\,d\theta \geq \frac{\sqrt{3}}{\sqrt{10}}\left[\left|1-\frac{\sin 2\pi\delta}{\pi}\left[\frac{1}{2\delta}+\frac{1}{2\delta-1}-\frac{2}{3}\left[\frac{1}{2\delta+1}+\frac{1}{2\delta+2}+\frac{1}{2\delta+3}\right]\right]\right|\right.$$

$$\left.-\sum_{k\neq 0}\frac{|\sin 2\pi\delta n_k|}{\pi}\left|\frac{1}{2\delta n_k}+\frac{1}{2\delta n_k-1}-\frac{2}{3}\left[\frac{1}{2\delta n_k+1}+\frac{1}{2\delta n_k+2}+\frac{1}{2\delta n_k+3}\right]\right|\right]$$

$$\geq \frac{\sqrt{3}}{\sqrt{10}}\left[\left|1-\frac{\sin 2\pi\delta}{\pi}\left[\frac{1}{2\delta}+\frac{1}{2\delta-1}-\frac{2}{3}\left[\frac{1}{2\delta+1}+\frac{1}{2\delta+2}+\frac{1}{2\delta+3}\right]\right]\right|\right.$$

$$\left.-\sum_{k\leq -4,\,k\geq 2}\frac{|\sin 2\pi\delta k|}{\pi}\left|\frac{1}{2\delta k}+\frac{1}{2\delta k-1}-\frac{2}{3}\left[\frac{1}{2\delta k+1}+\frac{1}{2\delta k+2}+\frac{1}{2\delta k+3}\right]\right|\right]$$

since $|n_{-k}|\geq 4$ for $k\geq 1$ and $n_k\geq 2$ for $k\geq 1$ if we assume that $|n_{-1}|\geq 4$. Note that $|n_{-1}|\geq 4$ also insures $2\delta n_k+i\neq 0$ for $i=0,-1,1,2,3$ and any $k$. □

Note that in (2) the negative $n_k$ in $P(\theta)$ satisfy $n_k-n_{k-1}\geq 2$ for $k<0$ and $n_{-1}\leq -2$. In stating the theorem below, we will be interested in finite sequences $(b_k)_{1\leq k\leq \ell}$ with the $b_k$ lying in between consecutive terms of $|n_k|$ for $k<0$. Note that such a sequence $b_k$ satisfies, $1\leq b_{k+1}-b_k\leq 2$. The following trivial lemma generates such sequences, whose proof is left to the reader.

**Lemma 9:** There are $2^{\ell-1}$ sequences $(b_k)_{1\leq k\leq \ell}$ with $b_1=1$ and $1\leq b_{k+1}-b_k\leq 2$. They may be generated by the following procedure: Given any $(\varepsilon_k)_{k=1}^{\ell-1}$ with $\varepsilon_k=0,1$ let the corresponding $(b_k)_{k=1}^{\ell}$ be $b_1=1$ and $b_{k+1}=b_k+1$ in case $\varepsilon_k=0$ and $b_{k+1}=b_k+2$ in case $\varepsilon_k=1$.

We may now get the estimate we want.

**Theorem 10:** Let $\varepsilon=-1$ and let $P(\theta)$ be as in (2). Then for any $\ell\geq 1$ and $\delta>\frac{1}{2}-\frac{1}{4\ell}$,

$$\frac{1}{2\delta}\int_{-\delta}^{\delta}|P(\theta)|^2 d\theta \geq \min\frac{1}{(2+4/\ell)^{1/2}}\left[\left|1-\frac{\sin 2\pi\delta}{\pi}\left[\frac{1}{2\delta}+\frac{1}{2\delta-1}-\frac{2}{\ell}\sum_{j=1}^{\ell}\frac{1}{2\delta+b_j}\right]\right|\right.$$

$$\left.-\sum_{\substack{|k|\geq 2n \\ k\notin\{-b_1,-b_2,\ldots,-b_\ell\}}}\frac{|\sin 2\pi\delta k|}{\pi}\left|\frac{1}{2\delta k}+\frac{1}{2\delta k-1}-\frac{2}{\ell}\sum_{j=1}^{\ell}\frac{1}{2\delta k+b_j}\right|\right]$$

where the minimum is over the $(b_k)_{k=1}^{\ell}$ in Lemma 9. Thus given $1\leq M<\sqrt{2}$ there is a $\delta_0<1/2$ such that for $\delta_0 \le \delta \le 1/2$,

$$\frac{1}{2\delta}\int_{-\delta}^{\delta} |P(\theta)|^2\, d\theta \ge M .$$

**Proof:** As in the proof of Theorem 8 the series in question converge uniformly. Let $P(\theta)$ be as in Theorem 2. Let $(b_k)_{k=1}^{\ell}$ be such that the $b_k$ lie in between the consecutive terms of $|n_k|$ for $k < 0$ with $b_1 = 1$, $1 \le b_{k+1} - b_k \le 2$. In the estimate of Theorem 6 set $a_0 = 1$, $a_1 = 1$, $a_{-k} = -2/\ell$ for $k = b_1, \ldots, b_\ell$ and all other $a_n = 0$. We get:

$$\frac{1}{2\delta}\int_{-\delta}^{\delta} |P(\theta)|^2 d\theta \ge \frac{1}{(2+4/\ell)^{1/2}}\left(\left|1-\frac{\sin 2\pi\delta}{\pi}\left[\frac{1}{2\delta}+\frac{1}{2\delta-1}-\frac{2}{\ell}\sum_{j=1}^{\ell}\frac{1}{2\delta+b_j}\right]\right|\right.$$
$$\left. -\sum_{k\neq 0}\frac{|\sin 2\pi\delta n_k|}{\pi}\left|\frac{1}{2\delta n_k}+\frac{1}{2\delta n_k-1}-\frac{2}{\ell}\sum_{j=1}^{\ell}\frac{1}{2\delta n_k+b_j}\right|\right)$$
$$\ge \frac{1}{(2+4/\ell)^{1/2}}\left(\left|1-\frac{\sin 2\pi\delta}{\pi}\left[\frac{1}{2\delta}+\frac{1}{2\delta-1}-\frac{2}{\ell}\sum_{j=1}^{\ell}\frac{1}{2\delta+b_j}\right]\right|\right.$$
$$\left. -\sum_{\substack{|k|\ge 2\\ k\notin\{-b_1,\ldots,-b_\ell\}}}\frac{|\sin 2\pi\delta k|}{\pi}\left|\frac{1}{2\delta k}+\frac{1}{2\delta k-1}-\frac{2}{\ell}\sum_{j=1}^{\ell}\frac{1}{2\delta k+b_j}\right|\right)$$

since $n_k \ge 2$ for $k \ge 1$ and $n_k \le -2$ for $k \le -1$. Moreover note that none of the denominators $2\delta n_k + b_j$ are zero by the choice of $b_j$ for $\delta > 1/2 - 1/4\ell$. The last part of the statement of the theorem is obvious by continuity and by making $\ell$ sufficiently large. □

**Acknowledgements**

It is a pleasure to thank Jerry Foschini, B. Gopinath, Jean-Pierre Kahane, Henry Landau, Hugh Montgomery and Brent Smith for some stimulating conversations on the subject matter of this paper. I would particularly like to thank Mike Honig for numerous conversations and for carrying out the numerical calculations.

## References

[1] G. Forney, "Lower Bounds on Error Probability in the Presence of Large Intersymbol Interference", IEEE Trans. Com., COM-20, No. 1 (1972), pp. 76-77.

[2] G. Foschini, "Performance Bound for Maximum Likelihood Reception of Digital Data", IEEE Trans. Information Theory, IT-21 (1975), pp. 47-50.

[3] D. Hajela, "On Computing the Minimum Distance for Faster than Nyquist Signaling", The 1987 Symposium on Information Theory and its Applications (SITA '87), November 1987, Enoshima Island, Japan.

[4] A. Ingham, "Some Trigonometrical Inequalities With Applications to The Theory of Series", Mathematische Zeitschrift, Vol. 41 (1936), pp. 367-379.

[5] J. Mazo, "Faster Than Nyquist Signaling", Bell System Technical Journal, Vol. 54, No. 8, (1975), pp. 1451-1462.

[6] A. Wyner, "Upper Bound on Error Probability For Detection With Unbounded Intersymbol Interference", Bell System Technical Journal, Vol. 54, No. 7, (1975), pp. 1341-1351.

[7] A. Zygmund, "Trigonometric Series", Vol. 2, Cambridge University Press, 1977.

[8] G. Foschini, "Contrasting Performance of Faster Binary Signaling with QAM", AT&T Bell Labs. Tech. J. 63 (1984), pp. 1419-1445.

[9] D. Hajela, "Some New Results on Faster Than Nyquist Signaling", Proceedings of the Twenty-first Annual Conference on Information Sciences and Systems, John Hopkins University, March 1987, pp. 399-403.

# Some new applications of the large sieve

Adolf Hildebrand

## 1. Introduction

The large sieve is an important tool in Analytic Number Theory. Originally conceived by Linnik [13] in 1941, it has been further developed and brought to use by a number of authors, notably Rényi [14], Roth [15] and Bombieri [1]. It has received several striking applications on classical problems in prime number theory such as the Goldbach and twin primes problems and the distribution of primes in arithmetic progressions; see [2] for a survey.

The large sieve is usually stated in the form of an inequality for finite sequences of complex numbers. We shall use here the large sieve in its arithmetic version (see, e.g., [5, p. 105]).

$$(1.1)\qquad \sum_{p\le\sqrt{N}} \frac{1}{p} \sum_{a=0}^{p-1} \left| \frac{p}{N} \sum_{\substack{M<n\le M+N \\ n\equiv a \bmod p}} a_n - \frac{1}{N} \sum_{M<n\le M+N} a_n \right|^2 \le C\,\frac{1}{N} \sum_{M<n\le M+N} |a_n|^2 ,$$

where $M$ and $N\ge 0$ are integers, $a_n$ ( $n=M+1,\dots,M+N$ ) are arbitrary complex numbers, $p$ runs through all primes $\le\sqrt{N}$ and $C$ is an absolute constant. Roughly speaking, this inequality says that a general sequence $(a_n)$ is well distributed <u>on average</u> over arithmetic progressions with prime moduli.

In most applications the inequality (1.1) is used only in the case the coefficients $a_n$ are 0 or 1, i.e., one takes $a_n = 1$ if $n$ belongs to a given finite sequence of integers $A$ and $a_n = 0$ otherwise. If the set $A$ carries an arithmetical structure and the distribution of $A$ on some of the residue classes $a \bmod p$ is known, then (1.1) can be used to derive further information on this set. For example, if $A$ is the set of integers in the interval $[M+1, M+N]$, which are not divisible by any prime from a given set of primes $\mathcal{P}$, then for each $p \in \mathcal{P}$ with $p \leq \sqrt{N}$ the residue class $0 \bmod p$ contributes $|A|^2/pN^2$ to the left-hand side of (1.1), and we obtain an upper bound for $|A|$. This example is typical for the application of the large sieve as a "sieve", i.e., a device to obtain upper and lower bounds for the cardinality of a set of integers from which certain residue classes have been "sifted" out.

Here we consider a different type of application, where the sieve aspect plays only a minor role. We shall apply (1.1) with coefficients $a_n$, which are multiplicative in $n$ and of modulus $\leq 1$, but otherwise arbitrary. In this case, the distribution of $(a_n)$ on zero residue classes is given by the simple formula

$$(1.2) \qquad \sum_{\substack{M < n \leq M+N \\ n \equiv 0 \bmod p}} a_n = a_p \sum_{\frac{M}{p} < n \leq \frac{M+N}{p}} a_n + O\left(\frac{N}{p^2} + 1\right) .$$

The large sieve (1.1), when used in conjunction with (1.2), then yields information on the average behavior of $a_n$. It thus opens up a new approach to the theory of multiplicative arithmetic functions.

The purpose of this paper is to outline the method involved and to survey the results which have been obtained in this way in a series of recent papers ([7]-[11]).

## 2. Mean values of multiplicative functions

A central problem in the theory of multiplicative functions is to determine the asymptotic behavior of the means

$$m(x) = m_f(x) = \frac{1}{x} \sum_{n \leq x} f(n)$$

of a multiplicative function $f : \mathbb{N} \longrightarrow \mathbb{C}$. In 1967 Wirsing proved the following elegant result.

Theorem (Wirsing [16]): For any real-valued multiplicative function $f$ of modulus $\leq 1$ the limit

$$\lim_{x \to \infty} m_f(x) \tag{2.1}$$

exists and is equal to the product

$$\prod_p \left(1 - \frac{1}{p}\right)\left(1 + \sum_{m \geq 1} \frac{f(p^m)}{p^m}\right) . \tag{2.2}$$

Wirsing's original proof is elementary, but highly complicated. In this section we shall outline a much simpler proof based on the large sieve which has been given in [7].

Our argument is based on the estimate

$$\sum_{p \leq \sqrt{x}} \frac{1}{p} \left|m(x) - f(p)m\left(\frac{x}{p}\right)\right|^2 \ll 1 , \tag{2.3}$$

which holds uniformly for $x \geq 1$ and all multiplicative functions $f$ of modulus $\leq 1$. This is easily proved on taking $M = 0$, $N = [x]$ and $a_n = f(n)$ in (1.1) and noting (1.2). Note that for this purpose the large sieve (1.1) is not needed in its full strength; for example, the contribution of the non-zero residue classes to the sum in (1.1) is not exploited.

We remark that the estimate (2.3) is not new; in fact, several authors have used this and similar estimates in the proof of the necessity parts of some mean value theorems; see, e.g., [3], [4], [12]. Applications of this type, however, are fairly straightforward. For example, if $f$ is a multiplicative function of modulus $\leq 1$, for which the mean value (2.1) exists and is non-zero, then (2.3) implies the convergence of the series $\Sigma |1 - f(p)|^2 p^{-1}$.

Here we are dealing with a quite different and more subtle application of (2.3), namely to prove the existence of the limit (2.1) under certain conditions on $f$. We shall in fact use (2.3) to establish the following result, which easily implies Wirsing's theorem but is also of independent interest.

<u>Proposition</u> [7]: Uniformly for $3 \leq x \leq x' \leq x^{5/4}$ and all real-valued multiplicative functions $f$ of modulus $\leq 1$ we have

$$(2.4) \qquad m(x') = m(x) + O(R(x,x')),$$

where

$$R(x,x') = \left(\log \frac{\log x}{\log (2x'/x)}\right)^{-1/2}.$$

Wirsing's theorem that, under the above assumptions on $f$, the limit $\lim_{x \to \infty} m(x)$ exists and is equal to the product (2.2), is an immediate consequence of the Proposition and the well-known (and easy-to-prove) fact that, under the same hypotheses, the "logarithmic mean value"

$$\lim_{x \to \infty} \frac{1}{\log x} \sum_{n \leq x} \frac{f(n)}{n} = \lim_{x \to \infty} \frac{1}{\log x} \int_1^x m(x') \frac{dx'}{x'}$$

exists and is equal to the product (2.2).

As a second application, the Proposition can be used to extend the range of validity of character sum estimates [8]. Let $(\frac{n}{p})$ denote the

Legendre symbol modulo $p$ . By Burgess' well-known results we have

$$(2.5)\qquad \sum_{n \leq x} \left(\frac{n}{p}\right) = o(x) ,$$

as $p \to \infty$ , uniformly for $x \geq p^{1/4+\varepsilon}$ , where $\varepsilon$ is any fixed positive number. The Proposition implies that for $p^{1/4} \leq x \leq p^{1/4+\varepsilon}$ the averages

$$\frac{1}{x} \sum_{n \leq x} \left(\frac{n}{p}\right) \quad \text{and} \quad \frac{1}{p^{1/4+\varepsilon}} \sum_{n \leq p^{1/4+\varepsilon}} \left(\frac{n}{p}\right)$$

differ by a quantity which tends to zero, as $p \to \infty$ and $\varepsilon \to 0$ . Hence it follows that the relation (2.5) remains valid in the larger range $x \geq p^{1/4}$ . Note that for this application the uniformity in $f$ of the estimate (2.4) is essential.

We now outline a proof of (2.4). We assume first that $|f| = 1$ . By the triangle inequality we deduce from (2.3) the estimate

$$(2.6)\qquad \sum_{p \leq \sqrt{x}} \frac{1}{p} \left( |m(x)| - |m(\tfrac{x}{p})| \right)^2 \ll 1 ,$$

which is easier to deal with, since it does not involve the values $f(p)$ . Using partial summation and a mild form of the prime number theorem, it is straightforward to replace the sum in (2.6) by an integral, and we get

$$(2.7)\qquad \int_2^{\sqrt{x}} \left( |m(x)| - |m(\tfrac{x}{y})| \right)^2 \frac{dy}{y \log y} \ll 1 ,$$

or equivalently

$$(2.8)\qquad \int_{\sqrt{x}}^{x/2} \left( |m(x)| - |m(y)| \right)^2 \frac{dy}{y \log(x/y)} \ll 1 .$$

Suppose now that $3 \leq x \leq x' \leq x^{5/4}$ , and let $I = [\sqrt{x'}, x/2]$ . Applying (2.8) with $x$ and $x'$ , we obtain

$$\Big(|m(x)|-|m(x')|\Big)^2 R(x,x')^{-2} \ll \Big(|m(x)|-|m(x')|\Big)^2 \int_I \frac{dy}{y\log(x'/y)}$$

$$\leq 2\int_I \Big\{\Big(|m(x)|-|m(y)|\Big)^2 + \Big(|m(x')|-|m(y)|\Big)^2\Big\} \frac{dy}{y\log(x'/y)} \ll 1 \quad .$$

This implies

$$(2.4)' \qquad |m(x)| = |m(x')| + O(R(x,x')) \qquad (3 \leq x \leq x' \leq x^{5/4}) ,$$

which is the asserted estimate (2.4), but with $|m(x)|$ in place of $m(x)$. However, noting that by assumption $f$ and hence $m(\cdot)$ are real-valued functions and that

$$m(x') = m(x) + O(\tfrac{1}{x}) \qquad (x \leq x' \leq x+1) ,$$

it is easily seen that the seemingly weaker estimate (2.4)' is actually equivalent to (2.4).

We thus have proved (2.4) under the additional assumption that $|f| = 1$. The same argument works if $1-|f(p)|$ is small on average over $p$. If this is not the case, then, in view of the hypothesis $|f| \leq 1$, $|f(n)|$ must be small on average, in which case a trivial upper bound for $|m(x)|$ and $|m(x')|$ turns out to be sufficient for (2.4).

The outlined argument constitutes the simplest form of our method. In the following sections we shall discuss variants and extensions of this argument, which lead to generalizations of Wirsing's theorem in several directions.

## 3. The prime number theorem

Wirsing's theorem contains the prime number theorem in its equivalent form

$$(3.1) \qquad \sum_{n \le x} \mu(n) = o(x) \qquad (x \to \infty) \quad ,$$

where $\mu$ is the Moebius function. A proof of Wirsing's theorem, which does not depend on the prime number theorem, would therefore provide a new proof of the prime number theorem in the form (3.1).

In the proof of the preceding section, we did use the prime number theorem (namely in order to deduce (2.7) from (2.6)), as did Wirsing in his original proof. However, by a modification of the above argument, it is possible to avoid an appeal to the prime number theorem and use instead only elementary prime number estimates of Mertens' type along with an upper bound for primes in short intervals obtained from Selberg's sieve, see [9]. One thus gets a new elementary proof of the prime number theorem in the form (3.1), which is substantially different from the previously known proofs.

## 4. Short interval results

As a natural extension of the mean value problem for multiplicative functions, one may study the behavior of the short interval averages

$$\frac{1}{\phi(x)} \sum_{x - \phi(x) < n \le x} f(n)$$

under suitable growth conditions on the function $\phi(x)$. This problem has not received much attention in the literature, mainly because the previously known methods of proving mean value theorems do not readily generalize to the short interval situation. The large sieve method of section 2, however, can easily be adapted to obtain short interval mean value theorems, and in particular yields the following generalization of Wirsing's mean value theorem.

<u>Theorem</u> [10]: Suppose that $0 < \phi(x) \leq x$ and $\log \phi(x) \sim \log x$, as $x \to \infty$. Then the limit

$$(4.1) \qquad \lim_{x\to\infty} \frac{1}{\phi(x)} \sum_{x-\phi(x) < n \leq x} f(n)$$

exists for all real-valued multiplicative functions of modulus $\leq 1$ and is equal to the product (2.2).

By an example of Erdös (cf. [10]) there exists, for any $\varepsilon > 0$, a multiplicative function $f$ assuming only the values 0 and 1 and such that in the case $\phi(x) = x^{1-\varepsilon}$ the limit (4.1) does not exist. Thus the growth condition $\log \phi(x) \sim \log x$ in the theorem is best-possible.

The proof of the theorem follows closely the argument of section 2. For $0 < \delta \leq 1$ and $x \geq 1$ let

$$m(x;\delta) = m_f(x;\delta) = \frac{1}{\delta x} \sum_{x(1-\delta) < n \leq x} f(n) .$$

The estimate (2.3) then holds with $m(x;\delta)$ in place of $m(x)$ and the summation restricted to $p \leq \sqrt{\delta x}$. Under the condition $|\log \delta| = o(\log x)$ the subsequent argument in section 2 goes through and yields the oscillation condition

$$m(x^{1+o(1)};\delta) = m(x;\delta) + o(1) ,$$

from which the asserted result in the form

$$\lim_{x\to\infty} m\left(x;\frac{\phi(x)}{x}\right) = \lim_{x\to\infty} m(x) = \prod_p \left(1-\frac{1}{p}\right)\left(1+\sum_{m\geq 1}\frac{f(p^m)}{p^m}\right)$$

easily follows.

We remark that this proof requires deeper information on the distribution of the primes than the proof of Wirsing's theorem. While in section 2 for the deduction of (2.7) from (2.6) the prime number theorem with logarithmic error term was sufficient (and even this could be relaxed, cf. section 3), the corresponding step with $m(x;\delta)$ in place of $m(x)$ requires for small $\delta$ Hoheisel's prime number theorem (or a lower bound for primes in short intervals of similar quality). The reason for this difference is that $m(x;\delta)$ varies more rapidly than $m(x)$ ; for example, we have trivially $m(x+z) = m(x) + o(1)$ , whenever $z = o(x)$ , but $m(x+z;\delta) = m(x;\delta) + o(1)$ only for $z = o(\delta x)$ .

## 5. Complex-valued multiplicative functions

Wirsing's theorem has been generalized to complex-valued multiplicative functions by Halász, who proved

**Theorem** (Halász [6]): Let $f: \mathbb{N} \to \mathbb{C}$ be a multiplicative function of modulus $\leq 1$ . Then there exist constants $A$ and $\alpha$ and a function $L(u)$ satisfying

$$|L| = 1 , \quad \sup_{1 \leq t \leq 2} |L(tu)-L(u)| = o(1) \quad (u \to \infty) ,$$

such that

$$m(x) = Ax^{i\alpha}L(\log x) + o(1) \quad (x \to \infty) . \tag{5.1}$$

For the proof of this theorem, Halász devised a new analytic method, which greatly influenced the further development of the theory of multiplicative functions. Wirsing's elementary method apparently is not capable of yielding Halász' theorem. However, the large sieve method can be adapted to deal with complex-valued multiplicative functions and leads to a new proof of Halász' theorem in the above-stated form. Moreover, by the same method one can extend Halász' theorem to short intervals, see [10].

We briefly sketch a proof of Halász' theorem by this method. The main task is to prove an appropriate generalization of (2.4), namely a relation of the type

$$m(x') = \left(\frac{x'}{x}\right)^{i\alpha} m(x) + O(\tilde{R}(x,x')) \qquad (3 \le x \le x' \le x^{5/4}), \tag{5.2}$$

where $\alpha$ is a suitable real number depending on $x$ and $f$ (but independent of $x'$) and the error term satisfies $\tilde{R}(x,x') = o(1)$ if $x' = x^{1+o(1)}$. Using such a relation, it is not hard to prove the desired result (5.1).

Relation (5.2) can be deduced from the estimate (2.3), but the deduction here is more difficult than in the case of real-valued functions. The reason for this is that in the latter case it was sufficient to prove (2.4)', i.e., the desired relation (2.4) for $|m(x)|$ instead of $m(x)$, and we therefore could use the simpler estimate (2.6) instead of (2.3). For complex-valued multiplicative functions, this is no longer possible. While the relation (2.4)' remains valid for complex-valued functions, it does not imply (2.4) in general.

We therefore have to use (2.3) directly in order to deduce (5.2). It turns out that for this deduction the precise definition of $m(x)$ is rather immaterial. One can show that, under fairly general assumptions on a function $m(x)$, the validity of an estimate of the type (2.3) over a sufficiently large range for $x$ implies that $m(x)$ satisfies a relation of the type (5.2). The argument involved here is somewhat intricate and ultimately rests on an approximate form of Cauchy's functional equation.

6. <u>Multiplicative functions at consecutive integers</u>

In the applications of the previous sections, the large sieve (1.1) was used only with the residue classes 0 mod p. In general, one cannot take advantage of the contribution of the non-zero residue classes to the sum in (1.1), since for non-zero residue classes a formula like (1.2) is not available. In this section, however, we shall give an application, where the contribution of these residue classes can be effectively exploited.

Suppose that $f : \mathbb{N} \longrightarrow \{-1, 1\}$ is a completely multiplicative function assuming only the values $\pm 1$, and let

$$N(x) = N_f(x) = \sum_{\substack{n \leq x \\ f(n+1) = -f(n)}} 1$$

be the number of sign changes up to $x$ in the sequence $\{f(n)\}$. A plausible, but probably deep conjecture is that

$$\liminf_{x \to \infty} N(x)/x > 0 ,$$

provided $f \not\equiv 1$. As an approximation to this conjecture we have

<u>Theorem</u> [11]: Let $f$ and $N_f(x)$ be as above, and suppose that $f \not\equiv 1$. Then

$$\limsup_{x \to \infty} N_f(x)(\log \log x)^3/x > 0 .$$

The proof of this result makes full use of the large sieve in the form (1.1). Its main idea can be described as follows. Suppose $N(x)$ is small. Then $f(n) = f(n+1)$ for "most" $n \leq x$. This implies that for small integers $a$, say $0 \leq a \leq A$, the quantities

$$\sum_{\substack{n \leq x \\ n \equiv a \bmod p}} f(n) \qquad \text{and} \qquad \sum_{\substack{n \leq x \\ n \equiv 0 \bmod p}} f(n)$$

are nearly equal for most primes $p$. For each $a \leq A$ the contribution of the residue classes $a \bmod p$ to the left-hand side of (1.1) (with $M = 0$, $N = [x]$ and $a_n = f(n)$) is therefore approximately equal to the contribution of the class 0 mod p . Consequently, the estimates derived from (1.1) on using only the zero residue classes can be improved by a factor 1/A , under the assumption that N(x) is small enough. For sufficiently large A , we arrive at a contradiction and thus deduce a lower bound for N(x) .

## References


[1] E. Bombieri, On the large sieve. Mathematika 12 (1965), 201-225.

[2] E. Bombieri, Le grand crible dans la théorie analytique des nombres. Astérisque 18 (1974).

[3] H. Daboussi and H. Delange, On a theorem of P.D.T.A. Elliott on multiplicative functions. J. London Math. Soc. (2) 14 (1976), 345-356.

[4] P.D.T.A. Elliott, A mean value theorem for multiplicative functions. Proc. London Math. Soc. (3) 31 (1975), 418-438.

[5] P.D.T.A. Elliott, Arithmetic Functions and Integer Products. Springer, New York 1985.

[6] G. Halász, Über die Mittelwerte multiplikativer zahlentheoretischer Funktionen. Acta Math. Acad. Sci. Hung. 19 (1968), 365-403.

[7] A. Hildebrand, On Wirsing's mean value theorem for multiplicative functions. Bull. London Math. Soc. 18 (1986), 147-152.

[8] A. Hildebrand, A note on Burgess' character sum estimate. C.R. Acad. Sci. Canada VIII (1986), 35-37.

[9] A. Hildebrand, The prime number theorem via the large sieve. Mathematika 33 (1986), 23-30.

[10] A. Hildebrand, Multiplicative functions in short intervals. Canad. J. Math., to appear.

[11] A. Hildebrand, Multiplicative functions at consecutive integers. Math. Proc. Cambridge Philos. Soc., to appear.

[12] K.-H. Indlekofer, A mean-value theorem for multiplicative functions. Math. Z. 172 (1980), 255-271.

[13] Ju. V. Linnik, The large sieve. Dokl. Akad. Nauk SSSR 30 (1941), 292-294.

[14] A. Rényi, On the large sieve of Ju. V. Linnik. Compositio Math. 8 (1950), 68-75.

[15] K. F. Roth, On the large sieves of Linnik and Rényi. Mathematika 12 (1975), 1-9.

[16] E. Wirsing, Das asymptotische Verhalten von Summen über multiplikative Funktionen II. Acta Math. Acad. Sci. Hung. 18 (1967), 411-467.

School of Mathematics
The Institute for Advanced Study
Princeton, N.J. 08540

Department of Mathematics
University of Illinois
Urbana, IL 61801

# ELLIPTIC FIBERINGS OF KUMMER SURFACES

William L. Hoyt
Rutgers University
Department of Mathematics
New Brunswick, New Jersey 08903

## 1. Introduction

Let $X$ be the Kummer surface for a principally polarized Abelian surface $A$. As usual this means that $X$ is the minimal non-singular model for the quotient $A/\pm 1$ modulo the action $p \longrightarrow \pm p$.

The purpose of this lecture is to describe a fibering $X \longrightarrow \mathbb{P}_1$ such that the generic fiber $X_s$ is an elliptic curve and such that the quotient map $A \longrightarrow A/\pm 1$ determines a ramified double cover $Y_s \longrightarrow X_s$ with four points of ramification and with Prym variety isogenous to $A$.

The fibering $X \longrightarrow \mathbb{P}_1$ is defined either (case I) in terms of an equation

$$y^2 = x(x-a_1)(x-a_2)(x-a_3)(x-a_4)$$

for a hyperelliptic curve $C$ of genus 2 in case $A = \text{Jac}(C)$, or (case II) in terms of Legendre equations

$$y = x(x-1)(x-c_i), \quad i = 1,2,$$

for a pair of elliptic curves $E_i$ in case $A = E_1 \times E_2$.

In each case this fibering can be identified with the Neron model for a twisted Legendre equation

$$y^2 = (t-b_1)(t-b_2)(t-b_3)x(x-1)(x-t)$$

relative to a field of the form

$$\mathbb{C}(\sqrt{(t-b_2)/(t-b_3)})$$

with $t$ transcendental over $\mathbb{C}$ and with parameters $b_j$ depending algebraically on the $a_i$ or $c_i$.

The proof is an extension of an argument used in Shioda [25] for a product E×E involving a particular elliptic curve.

A proof for case II is contained in previous lectures [13] (for E×E with general E) and [14] (for general $E_1 \times E_2$).

Only case I, A = Jac(C), is considered in the remainder of this lecture.

In this case arguments suggested by D. Morrison and I. Dolgachev show that the natural map $X \longrightarrow A/\pm 1$ carries the four points of ramification of $Y_s \longrightarrow X_s$ to two singular points of $A/\pm 1$ which are ordinary double points on the image of $X_s$ Consequently results in Griffiths and Harris [9] seem(?) to imply that if $A/\pm 1$ is embedded as a quartic surface in $\mathbb{P}_3$ as in [9], then the fibering $X \longrightarrow \mathbb{P}_1$ corresponds to the pencil of plane sections of $A/\pm 1$ through this pair of singular points.

Similar fiberings have been defined in Haine [10] and Barth [2] and Shioda and Inose [26].

## 2. Outline

The following is an outline of the lecture, including statements of the main theorems ((viii) and (x) below):
Let C be the hyperelliptic curve of genus 2 defined (up to resolution of a singularity at $\infty$) by

(i) $$y^2 = x(x-a_1)(x-a_2)(x-a_3)(x-a_4)$$

with distinct $a_i \in \mathbb{C}-\{0\}$; let A be the Jacobian variety for C; and let X be the Kummer surface for A.

Preliminary results in §§3-5 concern subfields of the field

(ii) $$L = \mathbb{C}(x_1, y_1, x_2, y_2)$$

generated by a pair of independent generic solutions $(x_i, y_i)$, $i = 1,2$, of (i).

In §3 the fixed fields are determined for subgroups of a group $\langle\epsilon,\delta\rangle$ of automorphisms of $L$ generated by

(iii) $$(x_1,y_1,x_2,y_2)^{\epsilon} = (x_2,y_2,x_1,y_1) \text{ and}$$
$$(x_1,y_1,x_2,y_2)^{\delta} = (x_2,y_2,x_1,-y_1).$$

In §4 surfaces and rational maps are determined which correspond (up to choice of model) to fixed fields and inclusions in §3. In particular it is noted that the function field for $X$ (or $A/\pm 1$) can be identified with the field

(iv) $$\mathbb{C}(s,x_0,y_0) \text{ with}$$
$$(s,x_0,y_0) = (x_1x_2,x_1+x_2,y_1y_2),$$

and that the map $A \longrightarrow A/\pm 1$ is induced by the inclusion

(v) $$\mathbb{C}(s,x_0,y_1+y_2) \supset \mathbb{C}(s,x_0,y_0).$$

In §5 it is shown that $\mathbb{C}(s,x_0,y_0)$ in (iv) is an elliptic function field over $\mathbb{C}(s)$ with a defining equation

(vi) $$y_0^2 = a_1a_2a_3a_4s(x_0-u_1)(x_0-u_2)(x_0-u_3)(x_0-u_4) \text{ with}$$
$$u_i = a_i+s/a_i \text{ for } 1 \le i \le 4,$$

and that (vi) is equivalent over $\mathbb{C}(s)$ to a twisted Legendre equation

(vii) $$y^2 = (t-b_1)x(x-1)(x-t) \text{ with}$$
$$b_1 = (a_1-a_3)(a_2-a_4)/(a_1-a_4)(a_2-a_3) \text{ and}$$
$$t = (u_1-u_3)(u_2-u_4)/(u_1-u_4)(u_2-u_3).$$

In §6 relations between the $a_i$, $b_1$ and the points of ramification of the quadratic extension $\mathbb{C}(s)/\mathbb{C}(t)$ are determined and are used to define a second twisted Legendre equation which is not equivalent to (vii) over $\mathbb{C}(t)$ but which becomes equivalent over $\mathbb{C}(s)$.

In §7 a minimality argument yields:

**(viii) Theorem**: The function s determines a fibering $X \longrightarrow \mathbb{P}_1$ which can be identified (up to isomorphism of fiber spaces) with the Neron model relative to $\mathbb{C}(s)$ for either of the equivalent equations (vi) or (vii). In particular the elliptic curve defined by $\mathbb{C}(s,x_0,y_0)/\mathbb{C}(s)$ is (up to isomorphism over $\mathbb{C}(s)$) a generic fiber $X_s$ of $X \longrightarrow \mathbb{P}_1$.

In §8 it is noted that the Picard numbers of A and X and the Mordell-Weil rank of $X \longrightarrow \mathbb{P}_1$ are related as follows:

(ix)

| | | | | |
|---|---|---|---|---|
| $\rho(A)$ | 1 | 2 | 3 | 4 |
| $\rho(X)$ | 17 | 18 | 19 | 20 |
| $r(X/\mathbb{P}_1)$ | 1 | 2 | 3 | 4. |

In §9 ramification is determined for double covers of curves over $\mathbb{C}(s)$ which correspond to quadratic extensions of fixed fields in §3.

In §10 Riemann surfaces and bases for homology and 1-forms are determined for some of these curves.

In §11 corresponding Riemann and Prym matrices are computed and are used to prove the final assertion of the following:

**(x) Theorem**: The extension (v) determines a ramified double cover $Y_s \longrightarrow X_s$ of a generic fiber of $X \longrightarrow \mathbb{P}_1$ with four points of ramification. These four points correspond under the natural map $X \longrightarrow A/\pm 1$, to a pair of singular points on $A/\pm 1$ which are also ordinary double points on the image of $X_s$. Furthermore, the Prym variety for $Y_s \longrightarrow X_s$ is isogenous to A.

In §12 there is a brief discussion of some related results and problems.

## 3. Fixed fields for subgroups of $\langle\epsilon,\delta\rangle$

The automorphisms $\epsilon,\delta$ in 2(iii) generate a dihedral group $\langle\epsilon,\delta\rangle$ of automorphisms of the field L in 2(ii) with relations

$\epsilon^2 = \delta^4 = 1$ and $\delta\epsilon = \epsilon\delta^3$. The lattice of subgroups has the form

(i)

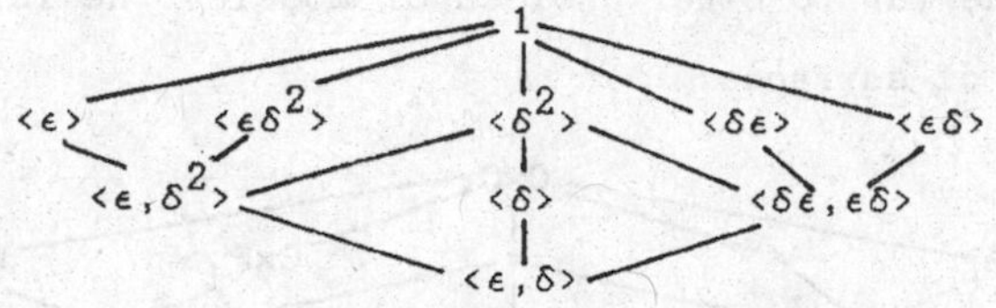


with each group having index 2 in each succeeding group.

The corresponding lattice of fixed fields has the form

(ii)

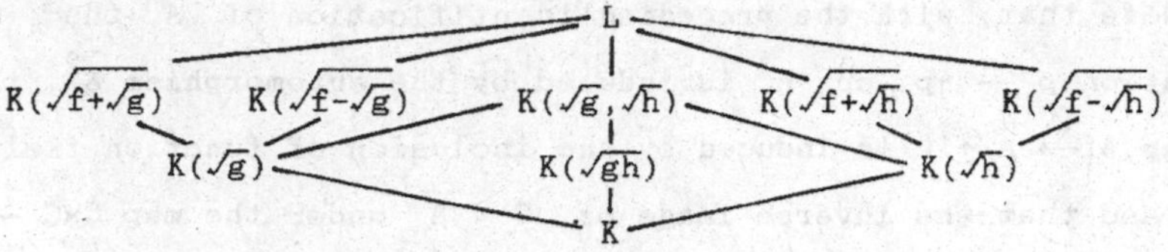


with $K$ and $f,g,h \in K$ defined by

(iii)
$$\begin{aligned} K &= \mathbb{C}(x_1x_2, x_1+x_2), \\ f &= y_1^2+y_2^2, \\ g &= 4y_1^2y_2^2, \\ h &= (y_1^2-y_2^2)^2, \end{aligned}$$

and with suitable labelling of $\sqrt{f\pm\sqrt{g}}$ and $\sqrt{f\pm\sqrt{h}}$. Note in particular that

$$\begin{aligned} & f^2 = g+h, \\ & (y_1^2-y_2^2)/(x_1-x_2) \in K, \\ & K(\sqrt{h}) = \mathbb{C}(x_1,x_2), \\ & K(\sqrt{f-\sqrt{h}}) = \mathbb{C}(x_1,x_2,y_2), \text{ and} \\ & [L:K] = 8. \end{aligned}$$

Also note that $L/K$ is the splitting field for

$$x^4 - 2fx^2 + g.$$

## 4. Corresponding rational maps of surfaces

If $A$ and $0 \in A$ are identified with the space of classes of positive divisors of degree two and with the class of special divisors of degree two, then the inclusions of function fields in 3(ii) determine (up to other choices of models) the following rational maps of surfaces:

(i)

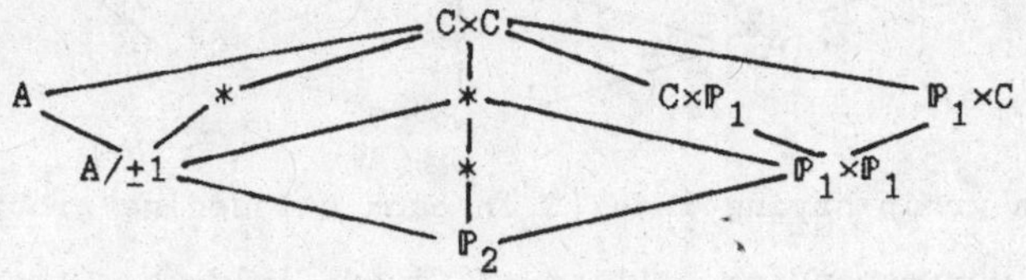


(ii) Note that, with the preceding identification of $A$, the involution $p \longrightarrow -p$ on $A$ is induced by the automorphism $\delta^2$, that the map $A \longrightarrow A/\pm 1$ is induced by the inclusion of function fields 2(v), and that the inverse image of $0 \in A$ under the map $C\times C \longrightarrow A$ is the transform $\Delta^{\delta}$ by $\delta$ of the diagonal $\Delta \subset C\times C$.

## 5. Equations for $\mathbb{C}(s,x_0,y_0)/\mathbb{C}(s)$

Let $a_i, b_1, x_0, y_0, s, u_i, t$ be as in 2(i)-(vii), and let

(i) $$x_0 \longrightarrow x = (u_1-u_3)(x_0-u_2)/(u_2-u_3)(x_0-u_1)$$

be the automorphism of $\mathbb{P}_1$ which maps $u_1,u_2,u_3,u_4$ to $\infty$, $0,1,t$, resp. Then routine substitution yields

(ii) $$(x_1-a_i)(x_2-a_i) = s-a_ix_o+a_i^2,$$

$$y_0^2 = s\Pi(a_i(x_0-u_i) \quad \text{(as required for 2(vi))},$$

$$u_i-u_j = (a_j-a_i)(s-a_ia_j)/a_ia_j,$$

$$t = b_1(s-a_1a_3)(s-a_2a_4)/(s-a_1a_4)(s-a_2a_3),$$

$$t-b_1 = -(a_1-a_2)(a_3-a_4)(a_1-a_3)(a_2-a_4)s/$$
$$a_1a_2a_3a_4(u_1-u_4)(u_2-u_3),$$

$$x-1 =(u_1-u_2)(x_1-u_3)/(u_2-u_3)(x_1-u_1),$$

$$x-t =(u_1-u_3)(u_1-u_2)(x_1-u_4)/(u_2-u_3)(u_1-u_4)(x_1-u_1),$$

and consequently

(iii) $(t-b_1)x(x-1)(x-t) = y^2$ with

$y = wy_0$ and

$w = ((a_2-a_1)(a_3-a_4)(a_1-a_3)(a_2-a_4))^{1/2}$

$(u_1-u_2)(u_1-u_3)/a_1a_2a_3a_4(u_2-u_3)^2(u_1-u_4)(x_0-u_1)$.

Therefore 2(vii) is equivalent over $\mathbb{C}(s)$ to 2(vi).

Results in the following §6 imply that there are distinct points $b_j \in \mathbb{P}_1 = \mathbb{C} \cup \{\infty\}$, $1 \le j \le 3$, such that

(iv) $\mathbb{C}(s) = \mathbb{C}(\sqrt{(t-b_2)/(t-b_3)})$

and such that the twisted Legendre equation

(v) $y^2 = (t-b_1)(t-b_2)(t-b_3)x(x-1)(x-t)$

is equivalent to 2(vii) over $\mathbb{C}(s)$ but not over $\mathbb{C}(t)$, provided that the factor $t-b_2$, resp. $t-b_3$ in (iv) and (v), is replaced by 1 in case $b_2 = \infty$, resp. $b_3 = \infty$.

## 6. Ramification for $\mathbb{C}(s)/\mathbb{C}(t)$

Note that $\{a_1a_3, a_2a_4\} \cap \{a_1a_4, a_2a_3\} = \emptyset$ since the $a_i$ are distinct and $\ne 0$. Therefore $t=t(s)$ in 5(ii) determines a double cover $\mathbb{P}_1 \to \mathbb{P}_1$ with $t(0) = t(\infty) = b_1$ and with ramification above two points $b_j = t(c_j)$, $c_j \ne 0,\infty$, $j = 2,3$. Also there are linear fractional functions $s_1$, resp. $t_1$, of $s$, resp. $t$, which satisfy

(i) $t_1 = s_1^2$

and which are determined by the correspondence between entries without parentheses in (ii) below with $c = c_3$, in which case it follows that $s_1$ and $t_1$ have the form specified in (iii) below and that there is a correspondence between entries with parentheses in (ii) with $d$ as in (iii):

| | | | | | | | |
|---|---|---|---|---|---|---|---|
| (ii) | $s$ | $(-c_3)$ | $+c_3$ | $0$ | $\infty$ | $(a_1a_3)$ | $(a_2a_4)$ |
| | $s_1$ | $(0)$ | $\infty$ | $-1$ | $+1$ | $(-d)$ | $(+d)$ |
| | $t$ | $b_2$ | $b_3$ | $b_1$ | $(b_1)$ | $(0)$ | $(0)$ |
| | $t_1$ | $0$ | $\infty$ | $1$ | $(1)$ | $(d^2)$ | $(d^2)$ |

(iii) $s_1 = (s+c)/(s-c)$,

$s = c(s_1+1)/(s_1-1)$,

$t_1 = (b_1-b_3)(t-b_2)/(b_1-b_2)(t-b_3)$,

$c = c_3 = -c_2$,

$c = (a_2a_4+c)/(a_1a_3-c)$,

$c_3^2 = (c(-d+1)/(-d-1))(c(+d+1)/(+d-1)) = (a_1a_3)(a_2a_4)$.

Conversely for each choice of distinct $b_j \in \mathbb{P}_1$, $1 \leq j \leq 3$, with $b_1 \neq 0,1,\infty$, there exist distinct $a_i \in \mathbb{C}-\{0\}$, $1 \leq i \leq 4$, such that

(iv) $b_1 = (a_1-a_3)(a_2-a_4)/(a_1-a_4)(a_2-a_3)$,

$b_2 = b_1(c-a_1a_3)(c-a_2a_4)/(c-a_1a_4)(c-a_2a_3)$ and

$b_3 = b_1(-c-a_1a_3)(-c-a_2a_4)/(-c-a_1a_4)(-c-a_2a_3)$ with

$c^2 = a_1a_2a_3a_4$.

These relations between the $a_i$ and $b_j$ determine an irreducible algebraic correspondence over $\mathbb{Q}$ which is generically 2-to-8:

(v) $T \subset \mathbb{P}_3 \times (\mathbb{P}_1)^3$.

## 7. The fibering $X \longrightarrow \mathbb{P}_1$

As in [4] and [5] it can be checked that the Neron model for 2(vii) relative to $\mathbb{C}(s)$ is an elliptic K3 surface which, by minimality of K3 surfaces, must be isomorphic to $X$. This proves 2(viii). Cf. Schmickler-Hirzebruch [23, p. 100].

## 8. Picard numbers and Mordell-Weil ranks

The singular fibers of $X \to \mathbb{P}_1$ correspond to the values $b_1, 0, 1, \infty$ for $t(s)$ and are of the following types, depending on the number of points in $\{b_2, b_3\} \cap \{0,1,\infty\}$:

$$2\ I_0^* \text{ and } 6\ I_2,$$
$$2\ I_0^*,\ 1\ I_4 \text{ and } 4\ I_2, \text{ or}$$
$$2\ I_0^* \text{ and } 2\ I_4 \text{ and } 2\ I_2 .$$

It can be checked that each $X \to \mathbb{P}_1$ has a nontorsion section corresponding to the value $b_1$ for $x$ in 2(vii), as well as the usual section "at $\infty$" and 3 sections of order 2 corresponding to the values $0, 1, t$ for $x$.

The relations listed in 2(ix) follow from well-known relations between Picard numbers $\rho(A)$, $\rho(X)$ for Abelian surfaces and their Kummer surfaces and between Picard numbers, Mordell-Weil ranks and components of singular fibers for elliptic surfaces. Cf. [24, 26].

## 9. Curves and double covers over $\mathbb{C}(s)$

(i) The lattice of quadratic extensions in 3(ii) determines the following lattice of ramified double covers of curves over $\mathbb{C}(s)$:

(i')

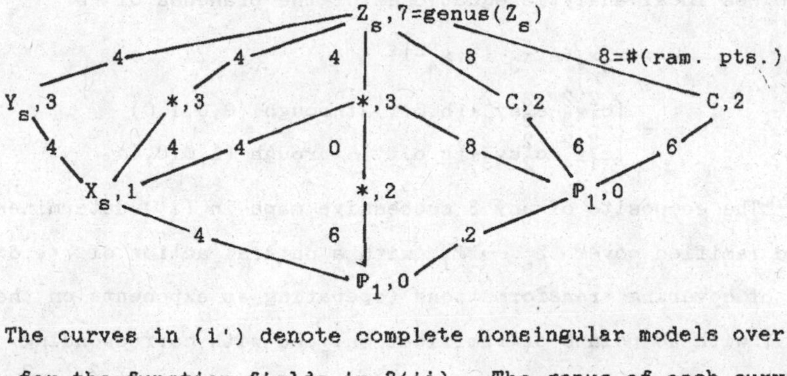


The curves in (i') denote complete nonsingular models over $\mathbb{C}(s)$ for the function fields in 3(ii). The genus of each curve and the number of points of ramification of each double cover are also specified in (i').

The only complicated step in determining ramification and genus is the following:

(ii) The natural image of $Z_s$ on $C\times C$ is a curve $Z''$ with two ordinary double points $P_0\times P_\infty$, $P_\infty\times P_0$ which, together with branches through these two points, can be defined as follows: For almost any $c \in \mathbb{C}$ let

$$x_1 = cx_3/(x_3-1),\ x_2 = cx_4/(x_4-1),$$
$$y_1 = y_3/(x_3-1)^3,\ y_2 = y_4/(x_4-1)^3 \text{ with}$$
$$y_k^2 = cx_k(x_k-1)\Pi(cx_k-a_ix_k-a_i) \quad \text{for} \quad k = 3,4,$$

and let $C'\times C'$ and $Z'$ be the loci of $(x_3,y_3,x_4,y_4)$ over $\mathbb{C}$ and $\mathbb{C}(s)$, resp. Then the natural birational map $C\times C \longrightarrow C'\times C'$ is biholomorphic at $(0,0,1,0),(1,0,0,0)$ on $C'\times C'$ and at corresponding points $P_0\times P_\infty, P_\infty\times P_0 \in C\times C$. The functions $y_3,y_4$ are local analytic parameters at each of these points with

$$x_j = c_1^2y_j^2 + \text{(higher order terms) and}$$
$$x_k-1 = c_2^2y_k^2 + \text{(h.o.t.)}$$

for suitable $c_1,c_2 \in \mathbb{C}-\{0\}$ and an obvious permutation $\{j,k\}$ of $\{3,4\}$. Consequently the equation $x_1x_2 = s$ for $Z''$ on $C\times C$ determines local analytic equations for the branches of $Z'$

$$0 = x_3x_4-s(x_3-1)(x_4-1)$$
$$= \begin{cases} c_1^2y_3^2-c_2^2sy_4^2+\text{(h.o.t)} & \text{through } (0,0,1,0) \\ c_1^2y_4^2-c_2^2sy_3^2+\text{(h.o.t.)} & \text{through } (1,0,0,0). \end{cases}$$

(iii) The composite of any 3 successive maps in (i') determines an 8-fold ramified cover $Z_s \longrightarrow \mathbb{P}_1$ with a natural action of $\langle\epsilon,\delta\rangle$ as group of covering transformations (operating as exponents on the right), with 28 points of ramification, and with corresponding stabilizers, values of $(x_i,y_i)$ and/or branches of $c_1^2y_j^2 - c_2^2sy_k^2$ as follows:

(iii')

| Stabilizer | $x_0$ | $(x_1,y_1,x_2,y_2)$ | $(x_3,y_3,x_4,y_4)$ | Branch |
|---|---|---|---|---|
| $<\epsilon>$ | $\pm 2\sqrt{s}$ | $\pm(\sqrt{s},+v,\sqrt{s},+v)$ | - | - |
| $<\epsilon\delta^2>$ | $\pm 2\sqrt{s}$ | $\pm(\sqrt{s},+v,\sqrt{s},-v)$ | - | - |
| $<\delta\epsilon>$ | $a_i+s/a_i$ | $(a_i,0,s/a_i,w)$ | - | - |
| $<\epsilon\delta>$ | $a_i+s/a_i$ | $(s/a_i,w,a_i,0)$ | - | - |
| $<\delta^2>$ | | $P_0\times P_\infty$ | $(0,0,1,0)$ | $c_1y_3\pm c_2\sqrt{s}y_4$ |
| $<\delta^2>$ | | $P_\infty\times P_0$ | $(1,0,0,0)$ | $c_1y_4\pm c_2\sqrt{s}y_3$ |

(iv) Points of ramification and genera for intermediate covers are easily determined from the data in (iii). In particular $Y_s \to X_s$ is ramified at four points of $Y$, which correspond to eight points of $Z_s$ which are fixed by $\epsilon\delta^2$ or $\delta^2$.

(v) As indicated in Fig. I in §10, the four fixed points for $\epsilon\delta^2$ on $Z_s$ correspond to four points on $C\times C$ which are transversal points of intersection of $Z"$ and $\Delta^\delta$; these four points (and by 4(ii) each point in $\Delta^\delta$) map to $0 \in A$ which is an ordinary double point on the image of $Y_s$; and this point maps to a singular point on $A/\pm 1$ which is an ordinary double point on the image of $X_s$. Also (as above) the four fixed points for $\delta^2$ on $Z$ map to two ordinary double points on $Z" \subset C\times C$; these two points map to a single point on $A$ which is of order two and which is an ordinary double point on the image of $Y_s$; and this point maps to a second singular point on $A/\pm 1$ which is also an ordinary double point on the image of $X_s$.

(vi) Bases for holomorphic 1-forms on the curves in (i') are easily determined. In particular

$$\omega_1 = dx_0/y_0 \quad \text{and} \quad \varphi_2 = dx_1/y_1,\ \varphi_3 = x_1dx_1/y_1$$

are bases for $X_s$ and $C$, resp. Also

$$\omega_1,\ \omega_2 = \varphi_2 + \varphi_2^\epsilon\ ,\ \omega_3 = \varphi_3 + \varphi_3^\epsilon$$

are linearly independent, $\epsilon$-invariant and holomorphic on $Z_s$ and hence form a basis for holomorphic 1-forms on $Y_s$.

**10. Riemann surfaces and bases for homology**

After replacing the universal domain and various fields of definition by $\mathbb{C}$ and by suitable subfields of $\mathbb{C}$, resp., one can represent the curves in 9(i') by m-sheeted ramified covers of the $x_0$-sphere with sheets corresponding to the branches of a generator of the corresponding extension of function fields.

(i) In particular L/K has a generator of the form $y_1+cy_2$ with $c \in \mathbb{Q}$; and $Z_s$ can be represented by an 8-sheeted cover of the $x_0$-sphere with sheets corresponding to the branches of $y_1+cy_2$, with crossings between sheets determined by analytic continuation of these branches across suitable branch cuts, and with the Galois group $\langle\epsilon,\delta\rangle$ acting as a corresponding group of covering transformations which interchange sheets.

More precisely let

$$L_1 \cup L_2 \cup L_3 \cup L_4 \cup L_5 \cup L_6$$

be a piecewise smooth arc in the $x_0$-sphere with no self-crossings and with segments $L_i$ extending from $a_1+s/a_1$ to $a_2+s/a_2$ to $a_3+s/a_3$ to $a_4+s/a_4$ to $\infty$ to $-2\sqrt{s}$ to $+2\sqrt{s}$, as indicated in Fig II below; and let U be the complement of this arc.

Then there are single-valued holomorphic branches of $y_1$ and $y_2$ on U which can be defined as in (iii) below and which transform by analytic continuation across the $L_i$ as follows:

(i')

| Transform | $L_1$ | $L_2$ | $L_3$ | $L_4$ | $L_5$ | $L_6$ |
|---|---|---|---|---|---|---|
| of $y_1$ | $-y_1$ | $+y_1$ | $-y_1$ | $+y_1$ | $-y_1$ | $+y_2$ |
| of $y_2$ | $+y_2$ | $+y_2$ | $+y_2$ | $+y_2$ | $-y_2$ | $+y_1$ |
| of $Z_s$ | $\epsilon\delta$ | 1 | $\epsilon\delta$ | 1 | $\delta^2$ | $\epsilon$ |

In this case the 8 sheets representing $Z_s$ can be identified with 8 copies of U with suitably attached copies of the $L_i$ and with corresponding branches of $y_1+cy_2$, crossings between sheets and sheet interchanges as indicated in Fig. III below.

(ii) In Fig. III blank copies of the $L_i$ indicate trivial analytic continuation of the corresponding branch; otherwise an element of $\langle\epsilon,\delta\rangle$ beside each $L_i$ indicates the sheet reached by crossing $L_i$ from either side. This notation will be used to identify certain integrals in §11 below.

(iii) To define the branches of $y_1$ and $y_2$ required in (i) first note that the inverse image of $U$ under the map $x_0 = x_1+s/x_1$ has a component $U'$ which contains the points $a_i$ and $\infty$ in the $x_1$-sphere as in Fig. IV below.

Then the required branch of $y_1$ can be defined as the composite of a branch of $\sqrt{\Pi(x_1-a_i)}$ on $U'$ with the inverse of the restriction of $x_1+s/x_1$ to $U'$. The required branch of $y_2$ is the transform of the specified branch of $y_1$ by analytic continuation across $L_6$ and corresponds to a branch of $\sqrt{\Pi((s/x_1)-a_i)}$ on $U'$. At the points $a_i+s/a_i$ this branch of $y_1$ (resp. $y_2$) is ramified of order two (resp. holomorphic) since corresponding properties are valid for the branch of $\sqrt{\Pi(x_1-a_i)}$ (resp. $\sqrt{\Pi((s/x_1)-a_i)}$ at corresponding points $a_i$ in the $x_1$-sphere. The transform of the specified branch of $y_2$ by analytic continuation across $L_6$ coincides with the specified branch of $y_1$ since each branch of $\sqrt{\Pi(x_1-a_i)}$ is transformed into itself by analytic continuation around a small loop about $+\sqrt{s}$ in the $x_1$-sphere.

(iv) As in Fig. V below $Y_s$ can be represented by 4 copies of $U$ with suitably attached copies of the $L_i$ and with sheets and crossing between sheets corresponding to branches of $y_1+y_2$ and to analytic continuation of these branches across the $L_i$.

Note that the corresponding extension $K(y_1+y_2)/K$ is not a Galois extension and that analytic continuation across $L_6$ leaves

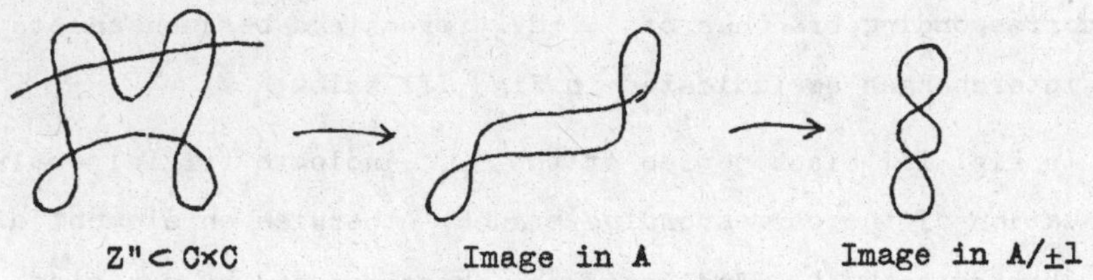


Figure I

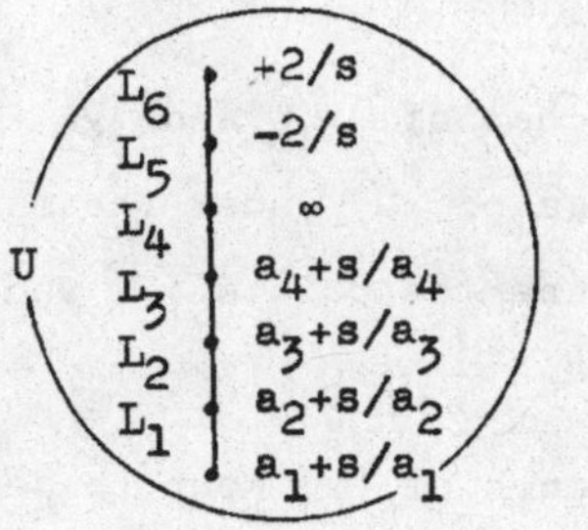


Figure II

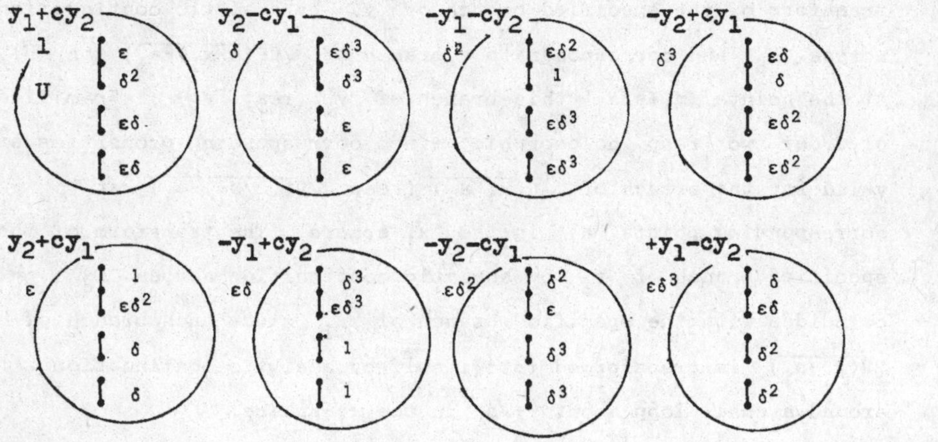


Figure III

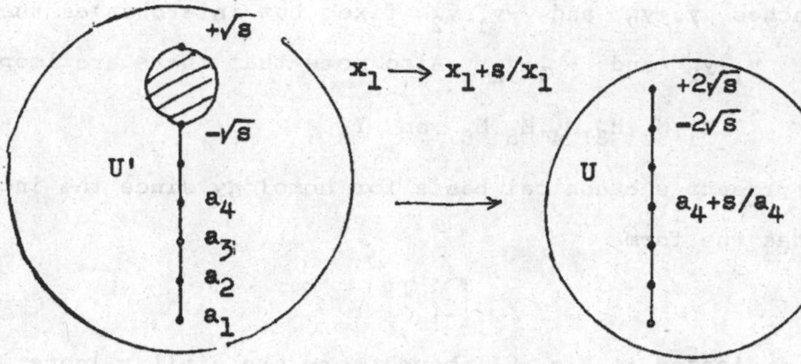


Figure IV

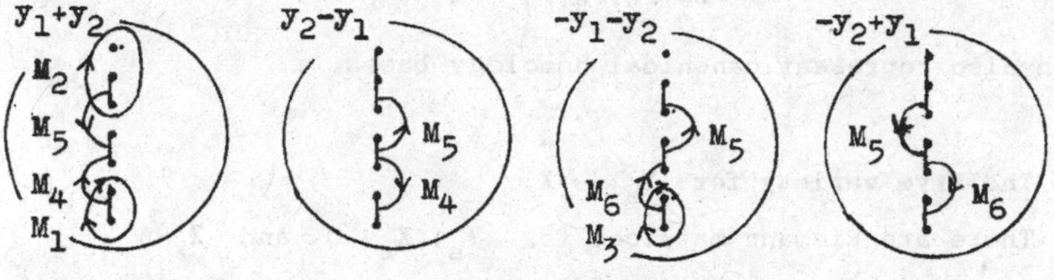


Figure V

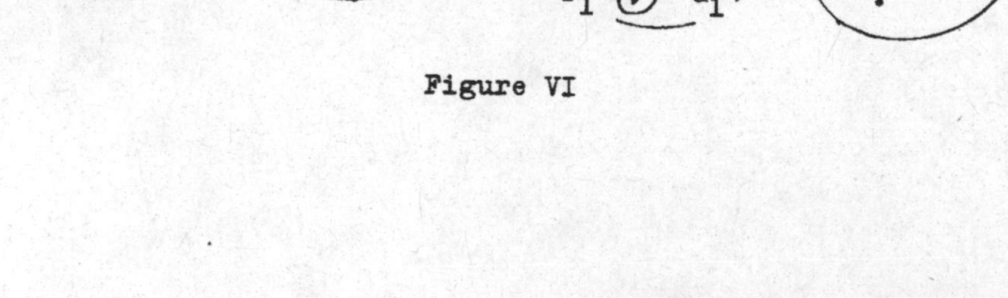

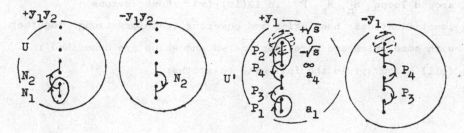

Figure VI

the branches $y_1+y_2$ and $-y_1-y_2$ fixed but interchanges the branches $y_1-y_2$ and $-y_1+y_2$. Also note that there are loops

(iv) $M_1, M_2, M_3, M_4, M_5, M_6$ on $Y_s$

which represent a canonical basis for homology since the incidence matrix has the form

$$\begin{bmatrix} 0 & I \\ -I & 0 \end{bmatrix}.$$

(v) As indicated in Fig. VI above there are similar loops

$$N_1, N_2 \text{ on } X_s \text{ and}$$

$$P_1, P_2, P_3, P_4 \text{ on } C$$

which also represent canonical homology bases.

## 11. The Prym variety for $Y_s \longrightarrow X_s$

There are Riemann matrices for $Y_s$, $X_s$, $C$ and $X_x \times A$

(i) $M = [m_{in}]$, $N = [n_{jp}]$, $P = [p_{kq}]$ and $Q = \begin{bmatrix} N & 0 \\ 0 & P \end{bmatrix}$

which are determined by integrals of 1-forms $\omega_i$, $\omega_j$, $\Psi_k$ in 9(vi) around loops $M_n$, $N_p$, $P_q$ in 10(iv)-(v). Some obvious identifications, homologies and covering transformations, together with some which are more complicated and which are described in (iii) below, yield the following relations:

(ii)
$$M = \begin{bmatrix} n_{11} & 0 & n_{11} & n_{12} & 0 & n_{12} \\ p_{21} & p_{22} & -p_{21} & p_{23} & 2p_{24} & -p_{23} \\ p_{31} & p_{32} & -p_{31} & p_{33} & 2p_{34} & -p_{33} \end{bmatrix}.$$

$$= Q\begin{bmatrix} 1 & 0 & 1 & 0 & 0 & 0 \\ 0 & 0 & 0 & 1 & 0 & 1 \\ 1 & 0 & -1 & 0 & 0 & 0 \\ 0 & 1 & 0 & 0 & 0 & 0 \\ 0 & 0 & 0 & 1 & 0 & -1 \\ 0 & 0 & 0 & 0 & 2 & 0 \end{bmatrix} \quad \text{and}$$

$$2Q = M\begin{bmatrix} 1 & 0 & 1 & 0 & 0 & 0 \\ 0 & 0 & 0 & 2 & 0 & 0 \\ 1 & 0 & -1 & 0 & 0 & 0 \\ 0 & 1 & 0 & 0 & 1 & 0 \\ 0 & 0 & 0 & 0 & 0 & 1 \\ 0 & 1 & 0 & 0 & -1 & 0 \end{bmatrix}$$

It follows from definitions in Fay [6] or Haine [10] that

$$A \cong (0\times\mathbb{C}^2+QZ^6)/QZ^6$$

is isogenous to

$$\mathrm{Prym}(Y_s/X_s) \cong (0\times\mathbb{C}^2+MZ^6)/MZ^6$$

as required in Theorem 2(x).

(iii) Note that

$$\omega_1^{\epsilon} = (dx_0/y_0)^{\epsilon} = +\omega_1 \quad \text{and} \quad \omega_1^{\delta} = \omega_1 \quad \text{since} \quad y_0 = y_1y_2.$$

After identifying suitable copies of the $L_i$ on $Y_s$, $Z_s$ and C, one obtains homologies

(iv)
$$M_5 \sim -L_4 + L_4^{\epsilon\delta} - L_4^{\epsilon\delta^3} + L_4^{\delta^2},$$

$$P_4 \sim -L_4 + L^{\epsilon\delta},$$

$$M_2 \sim L_5 + L_6 - L_6^{\epsilon} - L_5^{\delta^2},$$

$$P_2 \sim -L_5^{\delta^2} + L_5 + (-L_5^{\delta^2}+L_5)^{\epsilon}$$

which yield relations

(vi)

$$m_{15} = \int_{L_4} -\omega_1+\omega_1^{\epsilon\delta} - \omega_1^{\epsilon\delta^3} + \omega_1^{\delta^2} = 0,$$

$$m_{i5} = \int_{L_4} -\varphi_i-\varphi_i^{\epsilon} + (\varphi_i+\varphi_i^{\epsilon})^{\epsilon\delta} - (\varphi_i+\varphi_i^{\epsilon})^{\epsilon\delta^3} + (\varphi_i+\varphi_i^{\epsilon})^{\delta^2}$$

$$= -4\int_{L_4} \varphi_i$$

$$= 2p_{i5} \quad \text{for} \quad i = 2,3,$$

$$m_{12} = 0,$$

$$m_{i2} = \int_{L_5} \varphi_i+\varphi_i^{\epsilon} - (\varphi_i+\varphi_i^{\epsilon})^{\delta^2} + \int_{L_6} (\varphi_i+\varphi_i^{\epsilon}) - (\varphi_i+\varphi_i^{\epsilon})$$

$$= 2\int_{L_5} \varphi_i+\varphi_i^{\epsilon} = 2\int_{L_5} \varphi_i + 2\int_{L_5^{\epsilon}} \varphi_5$$

$$= 2p_{i2} \quad \text{for} \quad i = 2,3.$$

Similar but much simpler arguments for the other $m_{ij}$ yield the relation required for (ii).

(iv) Also note that

$$0\times\mathbb{C}^2+M\mathbb{Z}^6 = 0\times\mathbb{C}^2+Q\mathbb{Z}^6,$$

hence that $\mathrm{Prym}(Y_s/X_s)$ can be identified with the (connected) kernel of the map

$$\mathbb{C}^3/M\mathbb{Z}^6 \longrightarrow \mathbb{C}^3/(0\times\mathbb{C}^2+Q\mathbb{Z}^6)/Q\mathbb{Z}^6 \cong X_s,$$

also that there are isogenies of degrees 2 and 8

$$A \longrightarrow \mathrm{Prym}(Y_s/X_s) \longrightarrow A$$

which can be identified with the maps

$$(0\times\mathbb{C}^2+2Q\mathbb{Z}^6)/2Q\mathbb{Z}^6 \longrightarrow (0\times\mathbb{C}^2+M\mathbb{Z}^6)/M\mathbb{Z}^6 \longrightarrow (0\times\mathbb{C}^2+Q\mathbb{Z}^2)/Q\mathbb{Z}^2$$

with generators of kernels corresponding to

$$\begin{bmatrix}0\\p_{22}\\p_{32}\end{bmatrix} \text{ and } \begin{bmatrix}0\\p_{21}\\p_{31}\end{bmatrix}, \begin{bmatrix}0\\p_{23}\\p_{33}\end{bmatrix}, \begin{bmatrix}0\\p_{24}\\p_{34}\end{bmatrix}$$

in $M\mathbb{Z}^6/2Q\mathbb{Z}^6$ and $Q\mathbb{Z}^6/M\mathbb{Z}^6$, resp.

## 12. Related results and problems

Let $a_i$, $b_j$, $A$ and $X \longrightarrow \mathbb{P}_1$ correspond as in §§2,6,7 above. Also assume that $\vec{b} = {}^t(b_1,b_2,b_3)$ belongs to the set $U$ of all ${}^t(c_1,c_2,c_3) \in \mathbb{C}^3$ with distinct $c_i \neq 0,1,\infty$; and let $W \longrightarrow \mathbb{P}_1$ be the Neron model relative to $\mathbb{C}(t)$ for

(i) $$y^2 = (t-b_1)(t-b_2)(t-b_3)x(x-1)(x-t) \quad ((i) = 5(v)).$$

Then arguments in [12 and 13] can be used to show that $W \longrightarrow \mathbb{P}_1$ is an elliptic K3 surface with invariant

$$J = 4(t^2-t+1)/27t^2(t-1)^2;$$

with singular fibers of types

$$I_0^*, I_0^*, I_0^*, I_2, I_2, I_2 \quad \text{at} \quad b_1, b_2, b_3, 0, 1, \infty \in \mathbb{P}_1;$$

with Mordell-Weil rank

$$r = r(W/\mathbb{P}_1) = r(X/\mathbb{P}_1)-1 = \rho(A)-1$$

and group of holomorphic sections isomorphic to

$$(\mathbb{Z}/2\mathbb{Z})^2 \times \mathbb{Z}^r;$$

with a holomorphic 2-form $\omega$ and an associated many-valued modular form $\varphi$ satisfying

$$\omega = dt \wedge dx/y = c\varphi d\tau \wedge dz,$$
$$\varphi = (t(t-1)/(t-b_1)(t-b_2)(t-b_3))^{1/2}\Delta^{1/4},$$
$$t = \lambda(\tau) = \text{the Legendre function},$$
$$\Delta = \Delta(\tau) = \text{the Ramunujan function, and}$$
$$(\tau, z) = \text{suitable parameters on } W;$$

with a 5-dimensional parabolic cohomology group $H_{\mathbb{Z}}$, a nondegenerate quadratic form $q$ and a Hodge filtration

$$F^2 \subset F^1 \subset H_{\mathbb{Z}} \otimes \mathbb{C}$$

which are determined by periods and period relations for a vector-valued integral

$$\Phi = \int \varphi \begin{bmatrix} \tau \\ 1 \end{bmatrix} d\tau;$$

and with a 8-dimensional Abelian varity $A(H_{\mathbf{Z}})$ which is determined by an induced Hodge structure on the even part of the Clifford algebra for $(H_{\mathbf{Z}}, q)$.

Results of Shioda [24], Shioda and Inose [26] and Cox and Zucker [27] show that the Hodge structures on $H_{\mathbf{Z}}$, $H^2(W,\mathbf{Z})$, $H^2(X,\mathbf{Z})$ and $H^2(A,\mathbf{Z})$ have commensurable transcendental lattices; and results of Kuga and Satake [28], Oda [22] and Morrison [21] imply that $A(H_{\mathbf{Z}})$ is isogenous to the product $A^4$ of 4 copies of $A$. Therefore Th. 2(x) above provides a solution of the following problem for the special case of Kummer surfaces:

(ii) Determine whether Kuga-Satake varieties for general elliptic K3 surfaces can be described in terms of fixed parts of Prym varieties for suitable ramified covers of generic fibers.

This problem has been a primary motivation for the present work and also for [12,13,14].

I have also become very much interested in the following problems concerning $W \longrightarrow \mathbb{P}_1$.:

(iii) Express periods and period relations of $\omega$ and $\Phi$ explicitly in terms of one another. This is done implicitly in [27].

(iv) Determine $H_{\mathbf{Z}}$ and $q$ explicitly. This is done in [12,13] for somewhat simpler examples.

(v) Determine the variation (resp. degeneration) of Hodge structure on $H_{\mathbf{Z}}$ as $\vec{b}$ varies on $U$ (resp. as $b_j \longrightarrow 0, 1, \infty$ or $b_k$). It should (?) be possible to compute a corresponding monodromy representation

$$M': \pi_1(U) \longrightarrow SO(H_{\mathbf{Z}}, q)$$

explicitly and to construct an algebraic family $\{A(H_{\mathbf{Z}})\}$ over $U$. This has been done in [13] for a simpler example.

(iv) Find explicit values $\vec{b} \in U$ for which $r(W/\mathbb{P}_1) > 0$. It suffices to find $\vec{b}$ for which (i) has a $\mathbb{C}(t)$-rational solution $(x,y)$ with $y \neq 0$; or to find A corresponding to a point in some Humbert surface in the Siegel modular 3-fold. Cf. Franke [7], Hirzebruch and van der Geer [11], and Lee and Weintraub [17]. Related problems: find explicit equations for $\vec{b}$ which correspond to explicit equations listed in [11] for certain Humbert surfaces; find relations between endomorphisms of A and holomorphic sections of $W \longrightarrow \mathbb{P}_1$; in particular try to use endomorphisms of A to find generators for holomorphic sections of $W \longrightarrow \mathbb{P}_1$.

**References**


[1] M. Artin and H.P.F. Swinnerton-Dyer, The Shafarevich-Tate conjecture for pencils of elliptic curves on K3 surfaces, Inv. Math. **20**(1973), 249-266.

[2] W. Barth, Abelian Varieties with (1,2)-polarization, preprint.

[3] ________ and Hulek, Projective models of Shioda modular surfaces, Manuscr. Math **50** (1985), 73-132.

[4] P. Deligne, La conjectures de Weil pour les surfaces K3, Inv. Math. **12** (1975), 206-266.

[5] __________, Cycles de Hodge absolus et periodes des integrales des varietes Abeliennes, Soc. Math. Fr. Mem. nr **2** (1980), 23-33.

[6] J. Fay, Theta Functions on Riemann Surfaces, LNM 352, Springer-Verlag (1973).

[7] H.-G. Franke, Kurven in Hilbertsche Modulflachen und Humbertsche Flachen im Siegel-Raum, Bonner Math. Schriften Nr **104** (1978).

[8] G. van der Geer, On the geometry of a Siegel modular threefold, Math. Ann. **260** (1982), 317-350.

[9] P. Griffiths and J. Harris, *Principles of Algebraic Geometry*, John Wiley & Sons (1978).

[10] L. Haine, Geodesic flows on SO(4) and Abelian surfaces, Math. Ann. **263** (1983), 435-472.

[11] F. Hirzebruch and G. van der Geer, Lectures on Hilbert Modular Surfaces, Sem. Math. Sup. Univ. Montreal **77** (1981).

[12] W. Hoyt, On surfaces associated with an indefinite ternary lattice, in LNM 1135, Springer-Verlag (1985), 197-210.

[13] ______, Notes on elliptic K3 surfaces, in LNM 1240, Springer-Verlag (1987), 196-213.

[14] ______, Elliptic fiberings of Kummer surfaces for products, preprint.

[15] J.-I. Igusa, Arithmetic variety of moduli for genus two, Ann. Math. **72** (1960), 612-649.

[16] ________, On the graded ring of theta constants. I and II, Amer. J. Math **86** (1964), 219-246, and 88 (1966), 221-236.

[17] R. Lee and S. Weintraub, Cohomology of a Siegel modular variety of degree 2, in *Group Actions on Manifolds*, AMS Contemporary Math. **36** (1985), 433-488.

[18] ________________, Cohomology of $Sp_4(\mathbb{Z})$ and related groups and spaces, Topology **24** (1985), 391-410.

[19] ________________, On the transformation law for theta constants, Math. Gottingensis **36** (1985), 1-22.

[20] D. Morrison, On K3 surfaces with large Picard number, Inv. Math. **75** (1984), 105-121.

[21] ________, The Kuga-Satake variety of an Abelian surface, J. Alg. **92** (1985), 454-476.

[22] T. Oda, A note on the Tate conjecture for K3 surfaces, Proc. Japan Acad. **56** (1980), 296-300.

[23] U. Schmickler-Hirzebruch, Elliptische Flachen uber $\mathbb{P}_1\mathbb{C}$ und die hypergeometrische Differentialgleichung, Schriftenr. Math. Inst. Univ Munster **33** (1985).

[24] T. Shioda, On elliptic modular surfaces, JMS Japan **24** (1972), 20-59.

[25] _______, Algebraic cycles on certain K3 surfaces in characteristic p, in *Manifolds-Tokyo* 1973, Univ. Tokyo (1975), 357-364.

[26] ______ and H. Inose, On singular K3 surfaces, in *Complex Analysis and Algebraic Geometry*, Iwanami-Shoten (1977), 119-136.

[27] D. Cox and S. Zucker, Intersection numbers of sections of elliptic surfaces, Inv. Math **53** (1979), 1-44.

[28] M. Kuga and I. Satake, Abelian varieties attached to polarized K3 surfaces, Math. Ann. **169** (1967), 239-242.

[29] R. Livne, On certain covers of the universal elliptic curve, Thesis, Harvard (1981).

[30] U. Persson, Double sextics and singular K3 surfaces, in LNM 1124, Springer-Verlag (1983), 262-328.

# Recent Developments in the Theory of Rational Period Functions

by M.I. Knopp

I. Introduction. My interest in this subject began in the academic year 1956-57, when, at the suggestion of my teacher, Paul Bateman, I studied the dissertation of Hurwitz [10] in preparation for thesis work in the area of modular forms. I was particularly struck by Hurwitz's investigation of the series

(1.1) $$G_2(z) = \sum_{m,n \in Z}{}' \frac{1}{(mz+n)^2},$$

the Eisenstein series of weight 2 connected with the full modular group

(1.2) $$\Gamma(1) = \{\begin{pmatrix} a & b \\ c & d \end{pmatrix} | a,b,c,d, \in Z\ ,\ ad-bc = 1\}.$$

Hurwitz demonstrates that, in contrast to the Eisenstein series of higher weight,

$$G_{2k}(z) = \sum_{m,n \in Z}{}' (mz+n)^{-2k},\ k \geq 2,$$

which are modular forms of weight 2k on $\Gamma(1)$, $G_2$ is a kind of modular "quasi-form", of weight 2, satisfying the transformation equations

(1.3) $$G_2(z+1) = G_2(z),\ z^{-2}G_2(-\frac{1}{z}) = G_2(z) - \frac{2\pi i}{z},$$

for $z \in H = \{z = x+iy | y > 0\}$. Since $S = \begin{pmatrix} 1 & 1 \\ 0 & 1 \end{pmatrix}$ and $T = \begin{pmatrix} 0 & -1 \\ 1 & 0 \end{pmatrix}$ generate $\Gamma(1)$, (1.3) implies that $G_2$ has "reasonable" behavior under any transformation in (1). Hurwitz's work makes it plain that the appearance of the "period function" $(-2\pi i)/z$ in (1.3) is due to the conditional convergence of the series (1.1). (Functions with the functional equations (1.3) arise also as the logarithmic-derivatives of modular forms.)

Somewhat later, in my dissertation [11], I encountered a similar phenomenon, but this time the functions in question had negative weight and polynomial "periods' Specifically, in [11] I construct functions F analytic in $H$ such that

(1.4) $$F(z+1) = F(z),\ z^{2k}F(-\frac{1}{z}) = F(z) + p(z),$$

where $k \in Z^+$ and p(z) is a polynomial of degree at most 2k. We now recognize such F as "Eichler integrals", but at the time Eichler's classic work [3] was not yet known to me. (Either it had just appeared or it was about to appear.) Comparing

(1.2) and (1.3), I made the preliminary and tentative supposition that these in fact represented essentially all cases of functions exhibiting transformation equations with rational period functions under $\Gamma(1)$.

II. Modular integrals and the generalized Poincaré series. While temporarily at Tufts University in the summer of 1975 I focused seriously upon this hypothesis, in fact finding several proofs for it, all incorrect, as it turned out. Indeed, one of these "proofs" finally evolved into a method for the construction of an entirely new class of rational period functions for $\Gamma(1)$.

Before describing these I give some definitions. Suppose f is meromorphic in $H$ and satisfies

$$f(z+1) = f(z),\ z^{-2k}f(-\tfrac{1}{z}) = f(z) + q(z), \tag{2.1}$$

where $k \in \mathbb{Z}$ and q(z) is a rational function. Then we call f a modular integral (MI) on $\Gamma(1)$ of weight 2k, with rational period function (RPF) q. This definition can, of course, be generalized to odd integral weights, nontrivial multiplier systems and groups other than $\Gamma(1)$. (We shall have something to say later about RPF's on subgroups of $\Gamma(1)$.) Now, because $T^2 = (ST)^3 = I$ as linear functional transformations (the defining relations of $\Gamma(1)$), it follows from (2.1) that

$$z^{-2k}q(-\tfrac{1}{z})+q(z) = 0,\ (z-1)^{-2k}q(\tfrac{1}{z-1})+z^{-2k}q(\tfrac{z-1}{z})+q(z) = 0. \tag{2.2a}$$

Letting

$$F|\begin{pmatrix}a & b\\ c & d\end{pmatrix} = (cz+d)^{-2k}F(\tfrac{az+b}{cz+d}) \tag{2.3}$$

for $M = \begin{pmatrix}a & b\\ c & d\end{pmatrix} \in \Gamma(1)$ and F defined on $H$, we can rewrite (2.2a) as

$$q|T+q = 0,\ q|(ST)^2 + q|ST+q = 0, \tag{2.2b}$$

a direct consequence of (2.1), as we have indicated.

In fact, even more is true; (2.2) is equivalent to (2.1) in the following sense. Suppose q is a rational function (or, less restrictively, simply holomorphic in $H$ and of polynomial growth, both at $\infty$ and upon vertical approach to the real axis from within $H$) satisfying the relations (2.2). Then there exists f holomorphic in $H$ such that (2.1) holds. The proof of this involves the generalized Poincaré series of Eichler ([4], [5]).

We disgress briefly to describe the generalized Poincaré series, an often useful device which is easy to describe, yet apparently not widely known. For a discrete $\Gamma$ acting on $H$, the collection $\{q_M | M \in \Gamma\}$ is called a <u>cocycle in weight</u> 2k if

(2.4) $$q_{M_1 M_2} = q_{M_1}|M_2 + q_{M_2}, \text{ for } M_1, M_2 \in \Gamma,$$

where $q_{M_1}|M_2$ is defined by (2.3). To obtain a cocycle, we simply need to assign a $q_M$ to each M in a set of generators for $\Gamma(1)$ in such a way that the choice is consistent with the group relations among the generators. Then $q_M$ for general M in $\Gamma$ is defined by (2.4). That is, write M as a word in the generators and apply (2.4) repeatedly. In the case $\Gamma = \Gamma(1)$, the modular group, we can choose S and T as generators, with the defining relations $T^2 = (ST)^3 = I$. In particular, we wish to construct a cocycle $\{q_M\}$ such that $q_S = 0$, $q_T = q$, consistent with (2.1); then the conditions (2.2) on q are precisely the conditions of consistency with the two group relations in $\Gamma(1)$.

Now, given a rational function q satisfying (2.2) define the cocycle $\{q_M | M \in \Gamma(1)\}$ by application of (2.4) and form the <u>generalized Poincaré series</u>

(2.5) $$H(z) = \sum_{\substack{c,d \in Z \\ (c,d) = 1}} q_M(z)(cz+d)^{-2\rho},$$

where $M = \begin{pmatrix} * & * \\ c & d \end{pmatrix} \in \Gamma(1)$ and $\rho \in Z^+$, chosen sufficiently large to guarantee absolute convergence of the series. Note that $q_M$ depends only upon the lower row c,d of M as a consequence of $q_S = 0$. (There is a good deal of estimation required for the proof of absolute convergence; see [14, §II].) A function f satisfying (2.1) is then given by

(2.6) $$f(z) = -H(z)/E_{2\rho}(z),$$

where $E_{2\rho}$ is the Eisenstein series,

$$E_{2\rho}(z) = \sum_{\substack{c,d \in Z \\ (c,d) = 1}} (cz+d)^{-2\rho}.$$

($E_{2\rho}(z)$ is of weight $2\rho$ on $\Gamma(1)$; it is in fact virtually the same as $G_{2\rho}(z)$, defined above: $G_{2\rho}(z) = \zeta(2\rho)E_{2\rho}(z)$.) However, $f(z)$ defined by (2.6) falls short of what we require, since it may have poles at the zeros of $E_{2\rho}$. However, f can be modified to remove the poles. When $2k \geq 2$, we can accomplish this by application of a "Mittag-Leffler theorem" for automorphic forms; if $2k \leq 0$, elimination of the poles in $H$ (but not at $i\infty$) is still possible by use of a technically more complicated procedure based upon results of Douglas Niebur [16]. (See [14, §III] for details.) One should note that this construction, while it resolves the existence question for modular integrals, furnishes little insight into the relationship between the modular integral and its rational period function, or into the structure of the Fourier coefficients of the integral.

III. Examples of rational period functions. In the summer of 1975 I approached the problem of constructing RPF's, not through a search for modular integrals, but rather by focusing upon the relations (2.2). At that time I found the new class of examples

$$q_{2k}(z) = (z^2+z-1)^{-2k} + (z^2-z-1)^{-2k}, \tag{3.1}$$

with k an odd integer [12, Theorem 1]. My primary interest was in the case $k > 0$, in which instance $q_{2k}$ has poles of order k at the four real points $\pm(\sqrt{5}\pm1)/2$. (When $k \leq 0$, $q_{2k}$ is a polynomial of degree $-2k$, the period of an Eichler integral of weight 2k.) It is worth noting that these poles are the fixed points (and their negatives) of the commutator $STS^{-1}T^{-1}$. Further, the poles possess an algebraic symmetry; that is, if $z_0$ is a pole of $q_{2k}$, so is its algebraic conjugate.

While (3.1) represents only one example in each weight 2k, k odd, when $k > 0$ application of the usual Hecke operators $T_n (n \in Z^+)$ gives rise to a collection of infinitely many linearly independent RPF's in the same weight [12, §4]. To describe the action of $T_n$ on the RPF q, suppose q arises from the MI f. Then define

$$q^{(n)} = \hat{T}_n(q) = f_n|T - f_n,$$

where $f_n = T_n f$. Since we can show easily that $f_n$ is periodic, it follows that $q^{(n)}$ satisfies (2.2). Furthermore, q rational implies that $q^{(n)}$ is rational.

Now put $q_{2k}^{(n)} = \hat{T}_n q_{2k}$; the location of the poles of $q_{2k}^{(n)}$ shows that, with fixed k, these are all linearly independent. ($q_{2k}^{(n)}$ has 4n poles, all of them in $Q(\sqrt{5})$.)

The next nontrivial examples found of RPF's were given by Parson and Rosen [18], again of weight $2k \equiv 2 \pmod 4$, $k > 0$, and once again the poles were of order k and real quadratic irrationalities, this time lying in $Q(\sqrt{3})$ and $Q(\sqrt{21})$. (Furthermore, the poles again have the property of algebraic symmetry.) That this was not simply a coincidence is made clear by the following result concerning the poles of RPF's [13, Theorem 1].

Theorem. (i) If $z_0$ is a finite pole of a rational function q satisfying (2.2), then there is a squarefree $N \in Z^+$ such that $z_0 \in Q(\sqrt{N})$.
(ii) If the finite pole of q is in Q, then $z_0 = 0$.
Of course, this theorem makes a strong connection between RPF's and real quadratic fields, reinforcing the connection already suggested by (3.1). I should stress that part (i) of the theorem shows a pole of any RPF to be real, even if the corresponding MI has poles in $H$.

IV. Recent existence results. 1. My work [12, 13] and that of Parson/Rosen [18] left open a number of basic existence questions; namely:

(i) Do RPF's with irrational poles exist when the weight 2k is $\equiv 0 \pmod 4$?

(ii) For any given (squarefree) $N \in Z^+$, do RPF's exist with poles in $Q(\sqrt{N})$?

(iii) Do RPF's with irrational poles exist when the weight 2k is $\leq 0$?

(iv) Do there exist RPF's with irrational poles, but not exhibiting algebraic symmetry?

(v) Are there RPF's of weight 2k, with irrational poles of order other than k ?

(vi) Are there RPF's with irrational poles which are eigen-functions of the induced Hecke operators $\hat{T}_n$?

2. Thanks to work done during the past two years by Young-ju Choie [2] and John Hawkins [6], of this list now only (vi) remains open. (I conjecture that the

answer to (vi) is "no".) Choie has answered (i) and (ii) affirmatively, and in a strong sense. Specifically she has proved the

Theorem. Given $N \in Z^+$, N not a square, and any $k \in Z^+$, odd or even, there exist infinitely many linearly independent RPF's for $\Gamma(1)$ with poles in $Q(\sqrt{N})$ and no poles in Q (i.e. no poles at 0 or $\infty$).

Choie has devised several methods for the construction of nontrivial RPF's with irrational poles, that is, with poles at real quadratic irrationalities. One of these employs the "subgroup method", which entails constructing a cocycle of rational functions, in the sense of (2.4), on a suitably chosen subgroup $\Gamma$ of finite index in $\Gamma(1)$. Summation over the coset decomposition is then used to "lift" the cocycle on $\Gamma$ to one on the full group $\Gamma(1)$. The key factor in applying this method is the appropriate choice of the subgroup $\Gamma$.

If $\Gamma$ either is free or has a presentation with a single relation (as opposed to $\Gamma(1)$,which requires two relations), then the two conditions (2.2) are replaced by a single condition (for a single relation) or no condition (when the subgroup is free). A further requirement on the selection of $\Gamma$ is that it should admit a choice of cocycle which lifts to a nontrivial one on $\Gamma(1)$. For $\Gamma$ Choie selects the commutator subgroup of $\Gamma(1)$, a free group. This method appears to be restricted to weights $2k \equiv 2 \pmod 4$ and, while it gives RPF's corresponding to an infinite class of real quadratic fields $Q(\sqrt{N})$, it does not give RPF's for all of them.

Choie proves her theorem in full generality by means of another construction entirely. Making use of Pell's equation, this method reveals a further link with the theory of quadratic forms and quadratic fields. This link is further emphasized by the observation that all of the nontrivial examples of RPF's of weight 2k, k > 0, found by me, by Parson/Rosen, by Choie and, quite recently, by Parson [17] - as diverse as these are - have the form

(4.1) $$\sum \sigma(a,b,c)(az^2+bz+c)^{-k},$$

where the summation is over a finite set of $a,b,c \in Z$ and $\sigma$ is independent of z. Pursuing further the connection with quadratic forms, Choie has more recently exhibited an infinite class of quadratic fields with class number at least two [1]

3. The work of Hawkins appears to be only in its initial stages, but he has already produced examples of RPF's which answer question (iv) in the affirmative. He has, further, shown the answer to (iii) and (v) to be "no" in both cases. Furthermore, he gives the following example in weight 4:

$$(4.2)\qquad q(z) = \frac{8}{\sqrt{5}\,z} - \frac{2}{\sqrt{5}}\left\{\frac{2z+1}{(z^2+z-1)} + \frac{2z-1}{(z^2-z-1)}\right\} + \sqrt{5}\left\{\frac{2z+1}{(z^2+z-1)^2} + \frac{2z-1}{(z^2-z-1)^2}\right\}.$$

Clearly, $q(z)$ has the same four poles, $\pm(\sqrt{5}\pm 1)/2$, as do the RPF's $q_{2k}(z)$ given by (3.1), for $k$ odd. Now (4.2) obviously cannot be expressed as a linear combination of sums of the form (4.1), but, beyond this, Hawkins has shown that no RPF exists in weight 4 with these four poles and which is such a linear combination.

Hawkins formulates the notion of an "irreducible system" of poles belonging to any RPF of weight $2k > 0$ and, combining linear algebra and combinatorics, he shows that two distinct irreducible systems are disjoint. He proves, further, that there is at most one RPF (up to a constant multiple) associated to a given irreducible system, and he formulates the existence of this potential RPF in terms of the vanishing of $2k-3$ linear expressions in the (Eichler) polynomial periods of certain cusp forms of weight $2k$ on $\Gamma(1)$. (As it happens, these polynomial periods have been considered in detail in [15, §1].) But when $k = 1,2,3,4,5$ and 7 there are no nontrivial cusp forms of weight $2k$ on $\Gamma(1)$, as is well known. Thus the $2k-3$ homogeneous linear equations hold by default for these weights; that is, in weights 2,4,6,8,10 and 14 the RPF always exists corresponding to a given irreducible pole system.

Considering the number of poles in an irreducible system, Hawkins has found that the minimum number is four, and that, in this case, the poles are, in fact, $\pm(\sqrt{5}\pm 1)/2$. That is to say, when there are exactly four poles and $2k \equiv 2 \pmod 4$, $2k > 0$, the RPF's are precisely the $q_{2k}(z)$ given by (3.1). Hawkins' work proves existence as well for $2k = 4$ (see (4.2)) and $2k = 8$; however, he has also proved that <u>no</u> <u>RPF</u> <u>exists</u> with four poles when 2k = 12. It appears then that the cusp form

(4.3) $$\Delta(z) = e^{2\pi iz} \prod_{n=1}^{\infty} (1-e^{2\pi inz})^{24}$$

acts as an "obstruction" to the RPF for the irreducible system $\{\pm(\sqrt{5}\pm1)/2\}$ in weight 12, while the cusp form $E_6(z)\cdot\Delta(z)$ is <u>noc</u> an obstruction in weight 18. This, apparently because $18 \equiv 2 \pmod 4$. Clearly, much remains to be elucidated here.

V. <u>Modular integrals</u>. In [15, §2] Kohnen and Zagier discuss an interesting class of cusp forms on $\Gamma(1)$, introduced earlier by Zagier in [19]. These are defined by the "quadratic Eisenstein series,"

(5.1) $$f_{k,D}(z) = \sum (az^2+bz+c)^{-k},$$

where k is a positive even integer and $D \in Z^+$ is a discriminant; that is, $D \equiv 0$ or $1 \pmod 4$. The summation is over all triples $a,b,c \in Z$ such that $b^2-4ac = D$. $f_{k,D}(z)$ is a cusp form on $\Gamma(1)$ of weight $2k \equiv 0 \pmod 4$. (Note that $f_{k,D} = 0$ when k is odd.) In particular, [15] contains an explicit calculation of the even period polynomial $r^+(f_{k,D})$ arising from (2k-1)-fold integration of $f_{k,D}$.

To define $r^+(f_{k,D})$, we introduce the period polynomial r(f) associated to the cusp form f of weight 2k (and the element T of $\Gamma(1)$) by

(5.2) $$r(f)(X) = \int_{0=T(i\infty)}^{i\infty} f(z)(X-z)^{2k-2}dz.$$

One can show, without much difficulty, that such r(f) are RPF's of weight 2-2k, and, since $\Gamma(1)$ has the automorphism

$$\begin{pmatrix} a & b \\ c & d \end{pmatrix} \to \begin{pmatrix} a & -b \\ -c & d \end{pmatrix},$$

it follows that the polynomials $r^*(f)$ defined by $r^*(f)(X) = r(f)(-X)$ are again RPF's. The even and odd periods, $r^+(f)$ and $r^-(f)$ respectively, are then defined by

(5.3) $$r^+(f) = \frac{1}{2}\{r(f)+r^*(f)\}, \quad r^-(f) = \frac{1}{2}\{r(f)-r^*(f)\}.$$

Once again these are (polynomial) RPF's. By [15, Theorem 4], $r^+(f)$ can be expressed in the form

(5.4) $$r^+(f)(z) = \alpha\sum(az^2+bz+c)^{k-1} + \beta(z^{2k-2}-1),$$

where $\alpha$, $\beta$ are constants and the summation conditions in (5.4) are

(5.5) $$a,b,c \in Z, \; b^2-4ac = D, \; a < 0 < c.$$

The last condition in (5.5) guarantees that the sum in (5.4) is finite, hence that $r^+(f)$ is a polynomial of degree $\leq 2k-2$.

Both Choie and Hawkins have observed, independently, that the sums in (5.4) are RPF's, whether the (odd) exponent $k-1$ is positive or negative. When $k$ is even and $k \leq 0$ these sums are RPF's of the general type (4.1) and, in fact, if we put $D = 5$ and replace $k-1$ by $-k$, now with $k$ odd and $> 0$, they reduce to my early examples $q_{2k}(z)$ given by (3.1). Motivated by these observations, I carried out a formal $(2k-1)$-fold integration, term-by-term, of the series (5.1) for $f_{k,D}(z)$. While the integrated series clearly diverges, if we simply replace $k$ by $-k$ formally, we obtain the series

(5.6) $$\psi_{k,D}(z) = \sum \frac{\log((z-\beta)/(z-\alpha))}{(az^2+bz+c)^{k+1}},$$

where $k+1$ is odd and $> 0$. Here $\alpha = \dfrac{-b+\sqrt{D}}{2a}$, $\beta = \dfrac{-b-\sqrt{D}}{2a}$ and the summation is again over all triples $a,b,c \in Z$ with $b^2-4ac = D$ (as in (5.1)). When $k+1 \geq 3$, the series (5.6) converges absolutely; if $k+1 = 1$ the series can be handled by introduction (following Hecke) of a convergence factor (cf. [19, 39-42]). Note that when $k+1$ is even, $\psi_{k,D} = 0$.

Study of the function $\psi_{k,D}$, for $k+1$ odd, shows that it is holomorphic in $H$, periodic in $z$ with period 1 and that it satisfies the following transformation equation when subjected to the inversion $T$:

$$\psi_{k,D}|T - \psi_{k,D}$$

(5.7) $$= \sum_{b^2-4ac=D} \frac{\log(\alpha/\beta)}{(az^2+bz+c)^{k+1}} + 2\pi i \sum_{\substack{b^2-4ac=D \\ a<0<c}} (az^2+bz+c)^{-k-1}$$

$$= r_o(z) + r_e(z),$$

with summation again over $a,b,c \in Z$. The condition $a < 0 < c$ implies that $r_e(z)$ is a finite sum, and thus a rational function. Since $k > 0$, it has poles at the points

$\alpha, \beta \in Q(\sqrt{D})$. Further consideration shows that $r_e$ is an even function, while $r_o$ is odd. However, since the real line is a natural boundary for $r_o$, it is not possible to conclude from this alone (as we did above for the polynomials $r^+(f)$, $r^-(f)$ defined in (5.3)) that $r_o$ and $r_e$ satisfy (2.2). Notwithstanding this, these functions do in fact satisfy (2.2); for, as mentioned above, this has been verified directly for $r_e$, so it follows as well for $r_o$.

Thus, $\psi_{k,D}$ is a MI having even period the RPF $r_e$, but it would be more satisfactory to find, explicitly, a MI having $r_e$ as period (that is, without the additional odd period). Of course, the construction (2.6) involving the generalized Poincaré series is such a MI, but this is not explicit enough to permit calculation of its Fourier coefficients. On the other hand, the Fourier coefficients of a series similar to $\psi_{K,D}$, if one could be found with period $r_e$, presumably could be calculated (as I presume they can for $\psi_{k,D}$ itself) by the general method of Zagier [19, 44-45].

Beyond this there is the question of generalizing the construction to positive weights $2k \equiv 0 \pmod 4$.

VI. <u>The Mellin transform of a MI</u>. As is well known, Hecke, following Riemann, discovered - by applying the Mellin transform and its inverse - the systematic relationship between modular forms, on the one hand, and Dirichlet series with a simple functional equation, on the other [8,9]. In [13, Theorems 3 and 4], I showed that the same kind of bilateral relationship obtains between MI's with RPF's having poles in Q only (thus, at 0 and $\infty$ ) and a larger class of Dirichlet series with <u>precisely the same</u> functional equation as for the Mellin transform of a modular form. It follows, as a consequence of this relationship, that when the RPF of a MI has poles outside of Q (i.e. in $Q(\sqrt{N})$, with $N \neq$ a square), the Mellin transform of the MI cannot satisfy this same simple functional equation.

This observation serves as the starting point of my recent joint work with Hawkins [7], which in fact establishes a more complex functional equation for the Mellin transform of a MI with RPF, whether or not the poles of the RPF lie in Q. Compared with that of Hecke, this functional equation contains an additional term which is a finite sum of beta-functions, the number of terms depending upon the number of and orders of the poles of the RPF associated with the MI. We should note that while the ordinary Gaussian hypergeometric function ${}_2F_1$ figures prominently in the derivation of this functional equation, this function does not appear in the final expression for the additional term.

The derivation of the functional equation depends only upon the first (simpler) equation of (2.2), so that this derivation is applicable to the much wider class of RPF's on the subgroup $\Gamma_\theta$, generated by $S^2$ and T, which is of index 3 in $\Gamma(1)$. Presumably, there is an alternative approach that can make effective use of the full force of (2.2), to obtain a sharper form of the functional equation for the Mellin transform of a MI with RPF on $\Gamma(1)$. We have not yet found such an approach.

## References


1. Y. Choie, Rational period functions, class numbers and diophantine equations, preprint 19 pp.
2. Y. Choie, Rational period functions for the Hecke groups and real quadratic fields, preprint 47 pp.
3. M. Eichler, Eine Verallgemeinerung der Abelschen Integrale, Math. Z. 67 (1957), 267-298.
4. M. Eichler, Grenzkreisgruppen und kettenbruchartige Algorithmen, Acta Arith. 11 (1965), 169-180.
5. M. Eichler, Lectures on modular correspondences, Tata Institute of Fundamental Research, Bombay, 1955-56.
6. J. Hawkins, On rational period functions for the modular group, handwritten MS 113 pp.
7. J. Hawkins and M. Knopp, A Hecke correspondence theorem for automorphic integrals with rational period functions, preprint 71 pp.
8. E. Hecke, Lectures on Dirichlet series, modular functions and quadratic forms, Edwards Bros., Inc., Ann Arbor, 1938. Revised and reissued, Vandenhoeck & Ruprecht, Göttingen, 1983 (ed. B. Schoeneberg).

9. E. Hecke, Über die Bestimmung Dirichletscher Reihen durch ihre Funktionalgleichung , Math. Annalen 112 (1936), 664-699. Also, paper #33, pp. 591-626, in Mathematische Werke (ed. B. Schoeneberg), Vandenhoeck & Ruprecht, Göttingen, 1959.

10. A. Hurwitz,Grundlagen einer independenten Theorie der elliptschen Modulfunctionen und Theorie der Multiplicatorgleichungen erster Stufe, Math. Annalen 18 (1881), 528-591.

11. M. Knopp, Fourier series of automorphic forms of nonnegative dimension, Illinois J. Math. 5 (1961), 18-42.

12. M. Knopp, Rational period functions of the modular group, Duke Math. J. 45 (1978), 47-62.

13. M. Knopp, Rational period functions of the modular group II, Glasgow Math. J. 22 (1981), 185-197.

14. M. Knopp, Some new results on the Eichler cohomology of automorphic forms, Bull. Amer. Math. Soc. 80 (1974), 607-632.

15. W. Kohnen and D. Zagier, Modular forms with rational periods, Chapter 9, pp. 197-249, in Modular forms (ed. R. Rankin), Halsted Press, New York, 1984.

16. D. Niebur, Automorphic integrals of arbitrary positive dimension and Poincaré series, Doctoral Dissertation, University of Wisconsin, Madison, Wis., 1968.

17. L.A. Parson, Construction of rational period functions for the modular group, in preparation.

18. L.A. Parson and K. Rosen, Automorphic integrals and rational period functions for the Hecke groups, Illinois J. Math. 28 (1984), 383-396.

19. D. Zagier, Modular forms associated to real quadratic fields., Invent. Math. 30 (1975), 1-46.

Temple University
Philadelphia, PA.

## Additive problems in combinatorial number theory

Melvyn B. Nathanson
Office of the Provost and
Vice President for Academic Affairs
Lehman College (CUNY)
Bronx, New York 10468

### 1. Introduction

Many important questions in combinatorial number theory arise from the classical problems in additive number theory. Central to additive number theory is the study of bases of finite order. If A is a set of nonnegative integers, the <u>h-fold sumset</u> of A, denoted hA, is the set of all sums of h elements of A, with repetitions allowed. If hA is the set **N** of all nonnegative integers, then A is called a <u>basis of order h</u>. If hA contains all sufficiently large integers, then A is called an <u>asymptotic basis of order h</u>. Much of classical additive number theory is the study of sumsets hA, where A is the set of squares (Lagrange's theorem), or the k-th powers (Waring's problem), or the polygonal numbers (Gauss's theorem for triangular numbers and Cauchy's theorem for polygonal numbers of any order), or the primes (Goldbach's conjecture). Nathanson [24] recently gave a short and simple proof of Cauchy's polygonal number theorem.

Shnirel'man [32] created a new field of research in additive number theory when he discovered a powerful combinatorial criterion that implies that a set A is a basis of order h for some h, and proved that {0,1} U {primes} is a basis. Using Shnirel'man's method, Nathanson [23] proved that any set containing a positive proportion of the prime numbers is a basis of order h. Much of the work in combinatorial number theory concerns general properties of the classical additive bases and of arbitrary bases of finite order. In this paper I shall describe recent results and some unsolved additive problems in combinatorial number theory.

## 2. Thin bases

Let A be a set of nonnegative integers. Define the counting function $A(x)$ of the set A as the number of positive elements of A not exceeding x. If A is an asymptotic basis of order h, then an easy combinatorial argument shows that $A(x) > c\cdot x^{1/h}$ for some constant $c > 0$ and all x sufficiently large. An asymptotic basis A of order h is thin if $A(x) < c'x^{1/h}$ for some $c' > 0$ and all x sufficiently large. Thin bases exist. The first examples were constructed by Chatrovsky [2], Raikov [31], and Stöhr [33]. The best result is due to Cassels [1], who constructed, for each $h \geq 2$, a family of bases A of order h such that $A(x) \sim c\cdot x^{1/h}$ as x tends to infinity.

Problem 1. Do the classical sequences in additive number theory contain subsequences that are thin bases?

This is not yet known, but some surprising results have been obtained. Let A be a finite set of integers, and let $|A|$ denote the cardinality of A. If hA contains $\{0,1,2,\ldots,n\}$, then A is called a basis of order h for n. Clearly, if A is a basis of order h for n, then $|A| > n^{1/h}$. Choi, Erdös, and Nathanson [3] proved the following.

THEOREM 1. For every $n \geq 2$ there is a finite set A of squares such that A is a basis of order 4 for n, and

$$|A| < (4/\log 2)n^{1/3}\log n.$$

This was proved by means of an explicit construction. Note that there are $[n^{1/2}]+1$ squares up to n, and $|A|/n^{1/2}$ tends to zero.

Erdös and Nathanson [10] used probability methods to obtain the following result in the infinite case.

THEOREM 2. For every $\delta > 0$ there exists a set A of squares such that

(i) A is a basis of order 4,

(ii) If $n \neq 4^r(8k+7)$, then $n \in 3A$, and

(iii) $A(x) \sim c\cdot x^{(1/3)+\delta}$ for some $c > 0$.

Zöllner [38] combined the two results above to prove the following.

THEOREM 3. For every $\delta > 0$ there is an integer $n_0$ with the property that for any $n \geq n_0$ there is a finite set A of squares such that A is a basis of order 4 for n, and

$$|A| < n^{(1/4)+\delta}.$$

Problem 2. Does there exist $c > 0$ such that for $n \geq n_0$ there is a finite set A of squares such that A is a basis of order 4 for n, and

$$|A| < c \cdot n^{(1/4)} ?$$

Using probability methods, Zöllner and Wirsing independently obtained the following results on infinite sets of squares.

THEOREM 4 (Zöllner [37]). Let $h \geq 4$. For every $\delta > 0$ there exists a set A of squares such that A is a basis of order h and

$$A(x) < x^{(1/h)+\delta}$$

for x sufficiently large.

THEOREM 5 (Wirsing [36]). Let $h \geq 4$. There exists a set A of squares such that A is a basis of order h and

$$A(x) < c(x \log x)^{1/h}$$

for some $c > 0$ and all x sufficiently large.

Problem 3. Construct an explicit example of a set A of squares such that A is a basis of order 4 and $A(x)/x^{1/2}$ tends to zero. Note that this is considerably weaker that the non-constructive results stated in Theorems 2, 4, and 5.

Nathanson [21] has also obtained a thin variant of Waring's problem.

THEOREM 6. Let $k \geq 3$ and $s > s_0(k)$. Choose $\sigma$ such that $1-(1/s) < \sigma < 1$. There exists a set A of nonnegative k-th powers such that A is a basis of order s, and

$$A(x) \sim c \cdot x^{\sigma/k}$$

for some constant $c > 0$.

The proof requires probabilistic arguments as well as the Hardy-Littlewood asymptotic formula for the number of representations of an integer as the sum of s k-th powers.

Problem 4. Construct an explicit example of a set A of k-th powers such that, for some s, the set A is a basis of order s and $A(x)/x^{1/k}$ tends to zero.

There is a finite version of Theorem 6. Let f(n,k,s) denote the cardinality of the smallest finite set A of k-th powers such that A is a basis of order s for n. Clearly, $f(n,k,s) > n^{1/s}$. Define

$$\beta(k,s) = \limsup_{n \to \infty} \log f(n,k,s)/\log n.$$

Let g(k) denote the smallest integer h such that the set of all nonnegative k-th powers is a basis of order h. Nathanson [22] proved the following.

THEOREM 7. For $k \geq 3$ and $s \geq g(k)$,

$$f(n,k,s) < 2(s-g(k)+1)\cdot n^{1/(s-g(k)+k)}.$$

In particular, $\beta(k,s) \sim 1/s$ as s tends to infinity.

Finally, Wirsing [36] proved the following beautiful result on sums of primes.

THEOREM 8. For $h \geq 3$, there is a set P of prime numbers such that

(i) $n \in hP$ for all $n \equiv h \pmod 2$ and n sufficiently large,

(ii) $P(x) < c\cdot(x\log x)^{1/h}$.

## 3. Minimal bases

Recall that A is an <u>asymptotic basis of order h</u> if hA contains all sufficiently large integers. The asymptotic basis A is <u>minimal</u> if A is an asymptotic basis of order h, but no proper subset of A is an asymptotic basis of order h. This means that for each element $a \in A$ there are infinitely many positive integers n, each of whose representations as a sum of h elements of A must include the integer a as a summand. Stöhr [34] first defined minimal asymptotic bases, and the earliest results were obtained by Erdös, Härtter, and Nathanson [4,7,17,20].

It is important to note that not every asymptotic basis of

order h contains a subset that is a minimal asymptotic basis of order h. Stöhr [34], for example, observed that for $h \geq 2$ the set

$$A = \{1\} \cup \{ih \mid i = 0,1,2,\ldots\}$$

does not contain a minimal asymptotic basis of order h. For $h = 2$, Erdös and Nathanson [8] have constructed a set A with the property that, for any subset S of A, the set $A \setminus S$ is an asymptotic basis of order 2 if and only if S is finite. Since the infinite set A contains no maximal finite subset, it follows that A does not contain a minimal asymptotic basis of order 2.

Problem 5. Let $h \geq 3$. Construct a set A of nonnegative integers such that, for any subset S of A, the set $A \setminus S$ is an asymptotic basis of order h if and only if S is finite.

Erdös and Nathanson [9] have obtained a sufficient condition for an asymptotic basis of order 2 to contain a minimal asymptotic basis of order 2. For any set A of integers, let $r(n) = r_A(n)$ denote the number of solutions of the equation $n = a + a'$, where $a, a' \in A$ and $a \leq a'$.

THEOREM 9. Let A be an asymptotic basis of order 2. If $r(n) \geq c \cdot \log n$ for some constant $c > 1/\log(4/3)$ and all n sufficiently large, then A contains a minimal asymptotic basis of order 2.

Problem 6. Does the condition $r(n) > c \cdot \log n$ for <u>some</u> constant $c > 0$ and all n sufficiently large imply that A contains a minimal asymptotic basis of order 2?

In the opposite direction, Erdös and Nathanson [12] have proved the following.

THEOREM 10. Let $t \geq 1$. There exists an asymptotic basis A of order 2 such that $r(n) > t$ for all n sufficiently large, but A does not contain a minimal asymptotic basis of order 2.

Problem 7. Extend Theorems 9 and 10 to asymptotic bases of orders $h \geq 3$.

The following question seems to be very difficult.

Problem 8. If A is an asymptotic basis of order 2 such that r(n) tends to infinity, then must A contain a minimal asymptotic basis of order 2?

## 4. Fat minimal bases

Minimal bases are an extremal class of additive bases. I shall consider next an extremal property of this extremal class, namely, minimal asymptotic bases that are as "fat" or as "thin" as possible.

Recall that the <u>counting function</u> A(x) of the set A is the number of positive elements of A not exceeding x. Define the <u>lower asymptotic density</u> of A by $d_L(A) = \liminf A(x)/x$. If $\alpha = \lim A(x)/x$ exists, then $\alpha$ is called the <u>asymptotic density</u> of A, and denoted d(A).

Using Kneser's addition theorem [19] for the lower asymptotic density of sumsets, Nathanson and Sárközy [28] recently proved the following.

THEOREM 11. Let $h \geq 2$, and let A be an asymptotic basis of order h. If B is a subset of A such that $d_L(B) > 1/h$, then $A\backslash B$ contains a finite set F such that $B \cup F$ is an asymptotic basis of order h.

They drew two consequences from this result.

THEOREM 12. Let $h \geq 2$, and let A be an asymptotic basis of order h such that $d_L(A) = (1/h) + \delta$, where $\delta > 0$. Let $0 < \mu < \delta$. Then there is a subset W of A with asymptotic density $d(W) = \mu$ such that $A\backslash W$ is an asymptotic basis of order h.

THEOREM 13. Let $h \geq 2$. If A is a minimal asymptotic basis of order h, then $d_L(A) \leq 1/h$.

The next result shows that the estimate above is best possible.

THEOREM 14 (Erdös and Nathanson [11]). For every $h \geq 2$ there exist minimal asymptotic bases A of order h with $d(A) = 1/h$.

The proof of the theorem is by the somewhat complicated construction of explicit examples. The idea is as follows: Let A' be the union of sets B and C', where $B = \{b_i\}$ is a set of positive integers such that $b_i \equiv 1 \pmod h$ for all i, and $b_{i+1}/b_i$ tends to infinity, and where C' consists of all nonnegative multiples of h. Then A' is an asymptotic basis of order h with $d(A') = 1/h$. One can construct inductively a subset C of C' such that $C'\setminus C$ has density 0 and $A = B \cup C$ is a minimal asymptotic basis of order h. Then $d(A) = 1/h$.

It is worth noting that the preceding theorems are among the few results about minimal asymptotic bases that hold for all $h \geq 2$, and not just $h = 2$.

Problem 9. Let $h \geq 2$. Does there exist a minimal asymptotic basis A of order h such that $A(x) = x/h + O(1)$?

Examination of the proof of Theorem 14 in the case $h = 2$ shows that it produces a minimal asymptotic basis $A = \{a_i\}$ of order 2 such that $a_{i+1} > a_i$ and $\limsup (a_{i+1} - a_i) = 4$. It is easy to show that there cannot exist a minimal asymptotic basis $A = \{a_i\}$ of order 2 with $\limsup (a_{i+1} - a_i) = 2$.

Problem 10. Does there exist a minimal asymptotic basis $A = \{a_i\}$ of order 2 with $\limsup (a_{i+1} - a_i) = 3$?

For $h = 2$, Erdös and Nathanson [11] have extended Theorem 14 in the following way.

THEOREM 15. For every $\alpha \in (0, 1/2]$ there exists a minimal asymptotic basis A of order 2 with asymptotic density $d(A) = \alpha$.

Problem 11. Let $h \geq 3$. Show that for every $\alpha \in (0, 1/h]$ there exists a minimal asymptotic basis A of order h with asymptotic density $d(A) = \alpha$.

## 5. Thin minimal bases

The results in the preceding section are about minimal asymptotic bases that are as "fat" as possible. I shall now consider the construction of "thin" minimal asymptotic bases.

Recall that an asymptotic basis A is <u>thin</u> if $A(x) < c'x^{1/h}$ for some $c' > 0$ and all x sufficiently large. Nathanson [20] constructed the first example of a thin minimal asymptotic basis of order 2. This construction has recently been generalized to produce thin minimal asymptotic bases of order h for every $h \geq 2$.

THEOREM 16 (Nathanson [26]). Let $h \geq 2$. There exists a minimal asymptotic basis A of order h such that

$$c \cdot x^{1/h} < A(x) < c'x^{1/h}$$

for positive constants c and c' and all x sufficiently large.

Jia and Nathanson [18] have improved this as follows.

THEOREM 17. Let $h \geq 2$ and let $\mu \in [1/h, 1)$. There exists a minimal asymptotic basis A of order h such that

$$c \cdot x^{\mu} < A(x) < c'x^{\mu}$$

for positive constants c and c' and all x sufficiently large.

The minimal asymptotic bases in Theorems 16 and 17 are all constructed by the following method: If W is a subset of the nonnegative integers **N**, let A(W) consist of all numbers of the form $\Sigma_{f \in F} 2^f$, where F is a finite, nonempty subset of W. Let

$$\mathbf{N} = W_0 \cup W_1 \cup \cdots \cup W_{h-1}$$

be a partition of **N** into pairwise disjoint, nonempty sets. Then the set

$$A = A(W_0) \cup A(W_1) \cup \cdots \cup A(W_{h-1}) \qquad (*)$$

is an asymptotic basis of order h. If the partition is chosen appropriately, the basis A is minimal.

Not every partition, however, gives rise to a minimal basis. For $h = 2$, Nathanson [26] gave an example of a partition $\mathbf{N} = W_0 \cup W_1$ such that the corresponding asymptotic basis $A = A(W_0) \cup A(W_1)$ is not a minimal asymptotic basis of order 2.

Problem 12. Let $h \geq 2$. Determine the partitions of **N** into h sets $\mathbf{N} = W_0 \cup W_1 \cup \cdots \cup W_{h-1}$ such that the set A defined by (*) is <u>not</u> a minimal asymptotic basis of order h.

## 6. A multiplicative variant of the Erdös-Turán conjecture

Let $\mathbf{N}$ denote the set of nonnegative integers. Let $h \geq 2$, and let A be an asymptotic basis of order h. Let $r_h(n)$ denote the number of representations of n as the sum of h elements of A. In 1941, Erdös and Turán [14] conjectured that if A is an asymptotic basis of order 2, then $\limsup r_2(n) = \infty$. This has not yet been proven. More generally, one can conjecture that if A is an asymptotic basis of order h, then $\limsup r_h(n) = \infty$. The Erdös-Turán conjecture can be restated as follows: Let A be a subset of $\mathbf{N}$, and define $s = \liminf r_h(n)$ and $t = \limsup r_h(n)$. Then $s > 0$ implies that $t = \infty$.

Let $A_1,\ldots,A_h$ be subsets of $\mathbf{N}$. Let $r'(n)$ denote the number of representations of n in the form $n = a_1 +\cdots+ a_h$, where $a_i \in A_i$ for $i = 1,\ldots,h$. Define $s' = \liminf r'(n)$ and $t' = \limsup r'(n)$. The sets $A_1,\ldots,A_h$ form an <u>additive system of order h</u> if $s' > 0$, that is, if $r'(n) > 0$ for all n sufficiently large.

Here is a simple example of an additive system of order h: For $m \geq 2$, let $A_1$ consist of all nonnegative multiples of m, let $A_2$ consist of exactly k complete sets of residues modulo m, and let $A_i = \{0\}$ for $i = 3,\ldots,h$. Then the sets $A_1,\ldots,A_h$ form an additive system of order h such that $s' = t' = k$. This example shows that the analogue of the Erdös-Turán conjecture does not hold for additive systems.

It is remarkable that the multiplicative version of the Erdös-Turán conjecture is true. Let $\mathbf{N}^*$ denote the set of positive integers, and let B be a subset of $\mathbf{N}^*$. Let $h \geq 2$. If every sufficiently large integer can be represented as a product of h elements of B, with repetitions allowed, then B is called a <u>multiplicative asymptotic basis of order h</u>. Let $g(n)$ denote the number of representations of n as a product of h elements of B. Using results from extremal graph theory, Erdös [5] obtained the following result.

THEOREM 18. Let $h \geq 2$. If B is a multiplicative asymptotic basis of order h, then $\limsup g(n) = \infty$.

Recently, Nešetřil and Rödl [30] used Ramsey's theorem to give a short proof of this result.

Let $B_1,\ldots,B_h$ be subsets of $\mathbf{N}^*$. Let $g'(n)$ denote the number of representations of n in the form $n = b_1\cdots b_h$, where $b_i \in B_i$ for $i = 1,\ldots,h$. Define $s' = \liminf g'(n)$ and $t' = \limsup g'(n)$. The sets $B_1,\ldots,B_h$ form a <u>multiplicative system of order h</u> if $s' > 0$, that is, if $g'(n) > 0$ for all sufficiently large n.

Here is a simple example of a multiplicative system: Let $B_1 = \{1,2,4,8,...\}$ be the set of powers of 2, let $B_2 = \{1,3,5,7,...\}$ be the set of odd numbers, and let $B_i = \{1\}$ for $i = 3,...,h$. Since every positive integer n has a unique representation as a product $n = b_1 \cdots b_h$ with $b_i \in B_i$, the sets $B_1,...,B_h$ form a multiplicative system of order h with $s' = t' = 1$. Thus, $s' > 0$ does not imply that $t' = \infty$ for multiplicative systems.

Although this construction suggests that an analogue of the Erdös-Turán conjecture will not hold for multiplicative systems, the opposite, in fact, is true. Using a version of Ramsey's theorem, Nathanson [25] proved the following:

THEOREM 19. Let $B_1,...,B_h$ be a multiplicative system of order h. If $s' = \liminf g'(n) \geq 2$, then $t' = \limsup g'(n) = \infty$.

Indeed, Nathanson [25] obtained the following more precise result.

THEOREM 20. For $h \geq 2$, let $M(h)$ be the set of all pairs $(s',t')$ such that $s' = \liminf g'(n)$ and $t' = \limsup g'(n)$ for some multiplicative system $B_1,...,B_h$ of order h. Then

$$M(h) = \{(1,y) \mid y \in \mathbf{N}^*\} \cup \{(x,\infty) \mid x = 1,...,h\}.$$

Note that Theorem 20 implies Theorems 18 and 19.

Problem 13. Can Ramsey theory be applied to the additive Erdös-Turán conjecture?

## 7. Finite sumsets containing special sets

Let n be a positive integer, and let A be a subset of $\{1,2,...,n\}$. Denote the cardinality of A by $|A|$. Let $k \geq 3$ and $\delta > 0$. Szemerédi [35] proved that if $|A| > \delta n$ for $n \geq n(\delta,k)$, then A contains an arithmetic progression of length k. Nathanson and Sárközy [29] have obtained a lower bound for the length of the longest arithmetic progression contained in the h-fold sumset of a finite set.

THEOREM 21. Let N and k be positive integers. Let A be a subset of $\{1,2,\ldots,N\}$ such that

$$|A| \geq N/k + 1. \tag{3}$$

Then there exists an integer d with

$$1 \leq d \leq k-1 \tag{4}$$

such that if h and z are any positive integers satisfying the inequality

$$N/h + zd \leq |A| \tag{5}$$

then the sumset (2h)A contains an arithmetic progression with z terms and difference d.

Problem 14. Let $h \geq 2$ and let A be a "large" subset of $\{1,2,\ldots,n\}$. Find a good estimate for the length of the longest arithmetic progression contained in the sumset hA.

This problem is related to a question of Erdös and Freud [6]. They conjectured that if $A \subseteq \{1,2,\ldots,n\}$ and $|A| > n/3$, then there is a power of 2 that can be written as a sum of distinct elements of A. This is best possible, since, for $n = 3m$ and $A = \{3,6,9,\ldots,3m\}$, each sum of elements of A is divisible by 3, hence is not a power of 2. Using the method of trigonometric sums, Freiman [16] proved this conjecture. He showed that there is a constant $c > 0$ such that, for n sufficiently large, some power of 2 can be written as a sum of $c \cdot \log n$ distinct elements of A.

This result is not completely satisfactory, since the number of summands in Freiman's theorem tends to infinity as n tends to infinity. Does there exist an absolute constant h such that, for n sufficiently large, there is a power of 2 that can be represented as a sum of at most h distinct elements of A? Nathanson and Sárközy [29] have recently solved this problem. They used Theorem 21 to prove the following two results.

THEOREM 22. Let $m > 2^7 3^3 = 3{,}456$. If $A \subseteq \{1,2,\ldots,3m\}$ and $|A| \geq m+1$, then there is a power of 2 that can be written as the sum of at most 3,504 elements of A.

THEOREM 23. For m sufficiently large, if $A \subseteq \{1,2,\ldots,3m\}$ and $|A| \geq m+1$, then there is a power of 2 that can be written as the sum of at most 30,961 distinct elements of A.

Problem 15. Let $h_1$ (resp. $h_1'$) be the least integer such that, for m sufficiently large, if $A \quad \{1,2,\ldots,3m\}$ and $|A| > m$, then there is a power of 2 that can be written as a sum of exactly $h_1$ (resp. at most $h_1'$) distinct elements of A. Determine the values of $h_1$ and $h_1'$.

## 8. Infinite sumsets containing special sets

Let A be an infinite set of integers. Szemerédi's theorem implies that if $d_U(A) > 0$, then A contains arbitrarily long finite arithmetic progressions. The set A, however, does not necessarily contain an infinite arithmetic progression. Indeed, it is easy to construct an example of a set A with $d_U(A) = 1$ such that neither A nor any sumset hA contains an infinite arithmetic progression. For real numbers x and y, let $[x,y]$ denote the set of all integers n such that $x \le n \le y$. Let $\{t_n\}_{n=1}^{\infty}$ be a sequence of positive integers such that $t_{n+1}/t_n$ tends to infinity. Let

$$A = \bigcup_{n=1}^{\infty} [t_{2n} + 1, t_{2n+1}].$$

Then $d_U(A) = 1$, because

$$\begin{aligned} A(t_{2n+1})/t_{2n+1} &\ge (t_{2n+1} - t_{2n})/t_{2n+1} \\ &= 1 - t_{2n}/t_{2n+1} \to 1. \end{aligned}$$

Let $h \ge 1$. Since $hA \cap [ht_{2n-1} + 1, t_{2n}] = \emptyset$ for $n \ge n(h)$, the sumset hA contains arbitrarily long gaps, and so does not contain an infinite arithmetic progression.

Note that $d_L(A) = 0$ for the set A in the preceding example. If A is a set of nonnegative integers such that $d_L(A) > 0$, then some sumset hA must contain an infinite arithmetic progression. Denote by $h\hat{\ }A$ the set of all sums of h pairwise distinct elements of the set A. Erdös, Nathanson, and Sárközy [13] proved the following result.

THEOREM 24. Let A be a set of nonnegative integers such that $d_L(A) = \alpha \in (0,1/2]$. Let h be the smallest integer $\ge 1/\alpha$. Then

(i) $(h+1)\hat{\ }A$ contains an infinite arithmetic progression with difference $g \le h^2-h$, and

(ii) $(h^2-h)\hat{\ }A$ contains an infinite arithmetic progression with difference $g \le h+1$.

This result is best possible in the sense that for every $h \geq 1$ there exists a set A such that $d_L(A) = 1/h$, but the sumset hA does not contain an infinite arithmetic progression. For example, let $\{t_n\}$ be a strictly increasing sequence of positive integers such that $t_{n+1}/t_n$ tends to infinity, and let the set A be the union of the integers in the intervals $[t_{n-1}, (t_n/h) - \sqrt{t_n}]$. Then $d_L(A) = 1/h$ and $d_U(A) = 1$. Since the sumset hA is disjoint from the interval $(t_n - h\sqrt{t_n}, t_n)$ for all $n \geq n(h)$, it follows that hA contains arbitrarily long gaps, and so hA does not contain an infinite arithmetic progression.

Erdös, Nathanson, and Sárközy [13] have also proved the following result, which is an infinite analogue of the Erdös-Freud problem.

THEOREM 25. Let B be a set of nonnegative integers such that $d_L(B) \geq 1/3$ and $3 \nmid b^*$ for some $b^* \in B$. Then infinitely many powers of 2 can be written as sums of either four or five distinct elements of B.

Problem 16. Let $g_1$ (resp. $g_1'$) be the least integer such that, if A is any set of nonnegative integers with the properties that $d_L(A) \geq 1/3$ and $3 \nmid a$ for some $a \in A$, then some power of 2 can be written as the sum of exactly $g_1$ (resp. at most $g_1'$) elements of A. Determine the precise values of $g_1$ and $g_1'$.

In response to Theorem 25, Erdös and Freud [6] have posed the following problem.

Problem 17. Let A be a set of positive integers such that $d_L(A) > 1/3$. Does the equation $a_i + a_j = 2^t$ have infinitely many solutions with $a_i, a_j \in A$? If so, this result would be best possible.

## 9. Sumsets containing k-free numbers

There is an analogous problem concerning square-free numbers. Erdös and Freud [6] asked: If $A \subseteq \{1,2,\ldots,4m\}$ and $|A| \geq m+1$, then is there a square-free number that can be written as a sum of distinct elements of A? The set $A = \{4,8,12,\ldots,4m\}$ shows that this would be best possible. Nathanson and Sárközy (unpublished) obtained the following result.

THEOREM 26. For m sufficiently large, if $A \subseteq \{1,2,\ldots,4m\}$ and $|A| \geq m+1$, then there are at least $O(\sqrt{n})$ square-free numbers, each of which can be written as a sum of either 20 or 21 distinct elements of the set A.

Using a clever combinatorial argument, Filaseta [15] has greatly improved this result.

THEOREM 27. Let $A \subseteq \{1,2,\ldots,4m\}$ be of maximal cardinality such that

(i) $A \nsubseteq \{4,8,12,\ldots,4m\}$,

(ii) $A \nsubseteq \{2,6,10,\ldots,4m-2\}$,

(iii) 2A contains no square-free number.

Then

$$2/9 \leq \liminf |A|/m \leq \limsup |A|/m \leq 4-32/\pi^2 = 0.757\ldots.$$

Filaseta has asked if $\lim_{n\to\infty} |A|/m$ exists.

Let $Q_k$ denote the set of all k-free natural numbers, and let $Q_k'$ denote the set of all odd, k-free numbers. The set $Q_k$ has asymptotic density $1/\zeta(k)$, where $\zeta(k)$ is the Riemann zeta function, and $Q_k'$ has asymptotic density $2^{k-1}/((2^k-1)\zeta(k))$.

Define the subset sum s(B) by $s(B) = \Sigma_{b\in B} b$. It is easy to find sets A such that $s(B) \notin Q_k$ for all subsets $B \subseteq A$. For example, let A be a set of multiples of $d^k$ for some $d \geq 2$. Then $d^k | s(B)$ for all subsets B of A, and so $s(B) \notin Q_k$. Let $h \geq 2$. If we wish to consider only subset sums s(B) with $|B| = h$, then any set A, each of whose elements satisfies $a \equiv h^{k-1} \pmod{h^k}$, will have the property that $s(B) \notin Q_k$ whenever $B \subseteq A$ and $|B| = h$. In the case $h = 2$, if A is any subset of

$$\{n \geq 1 \mid n \equiv 2^{k-1} \text{ or } 2^{k-1}(3^k-1) \pmod{6^k}\},$$

then $a+a' \notin Q_k$ for all $a, a' \in A$. Nathanson [27] has given an upper bound for the size of any set $A \subseteq \{1,2,\ldots,n\}$ with the property that $a+a' \notin Q_k$ for all $a, a' \in A$.

THEOREM 28. Let $k \geq 2$ and $\delta > 0$. For n sufficiently large, if $A \subseteq \{1,2,\ldots,n\}$ satisfies the condition that $a+a' \notin Q_k$ for all a, a' $\in$ A with $a \neq a'$, then either

(1) $A \subseteq \{a \equiv 0 \pmod{2^k}\}$, or

(2) $A \subseteq \{a \equiv 2^{k-1} \pmod{2^k}\}$, or

(3) $|A| < n(1 - (2^k/((2^k-1)\zeta(k))) + \delta) < n/2^k$.

It follows from this result that if $A \subseteq \{1,2,\ldots,2^k m\}$ and $|A| \geq m+1$, then there exist a, a' $\in$ A with $a \neq a'$ and $a+a' \in Q_k$. Note that Filaseta's theorem is the case $k = 2$ of Theorem 28.

Problem 18. Let A be the largest subset of $\{1,2,\ldots,n\}$ such that $a+a' \notin Q_k$ for all a, a' $\in$ A with $a \neq a'$, and A is not of the form (1) or (2) in Theorem 28. Calculate lim sup $|A|/n$.

## REFERENCES

1. J. W. S. Cassels, Über Basen der natürlichen Zahlenreihe, Abh. Math. Sem. Univ. Hamburg 21 (1957), 247-257.

2. L. Chatrovsky, Sur les bases minimales de la suite des nombres naturels (Russian), Izv. Akad. Nauk SSSR Ser. Mat. 4 (1940), 335-340.

3. S. L. G. Choi, P. Erdös, and M. B. Nathanson, Lagrange's theorem with $N^{1/3}$ squares, Proc. of the Amer. Math. Soc. 79 (1980), 203-205.

4. P. Erdös, Einige Bemerkungen zur Arbeit von A. Stöhr "Gelöste und ungelöste Fragen über Basen der natürlichen Zahlenreihe," J. reine angew. Math. 197 (1957), 216-219.

5. P. Erdös, On the multiplicative representations of integers, Israel J. Math. 2 (1964), 251-261.

6. P. Erdös and R. Freud, personal communication.

7. P. Erdös and E. Härtter, Konstruktion von nichtperiodischen Minimalbasen mit der dichte 1/2 für die Menge der nichtnegativen ganzen Zahlen, J. reine angew. Math. 221 (1966), 44-47.

8. P. Erdös and M. B. Nathanson, Oscillations of bases for the natural numbers, Proc. Amer. Math. Soc. 53 (1975), 253-258.

9. P. Erdös and M. B. Nathanson, Systems of distinct representatives and minimal bases in additive number theory, in: M. B. Nathanson, ed., Number Theory, Carbondale 1979, Lecture Notes in

Mathematics, vol. 751, Springer-Verlag, Berlin-New York, 1979, pp. 89-107.

10. P. Erdös and M. B. Nathanson, Lagrange's theorem and thin subsequences of squares, in: J. Gani and V. K. Rohatgi (eds.), Contributions to Probability, Academic Press, New York, 1981, pp. 3-9.

11. P. Erdös and M. B. Nathanson, Minimal asymptotic bases with prescribed densities, Illinois J. Math. 32 (1988), 562-574.

12. P. Erdös and M. B. Nathanson, Asymptotic bases with many representations, Acta Arith., to appear.

13. P. Erdös, M. B. Nathanson, and A. Sárközy, Sumsets containing infinite arithmetic progressions, J. Number Theory 28 (1988), 159-166.

14. P. Erdös and P. Turán, On a problem of Sidon in additive number theory and some related questions, J. London Math. Soc. 16 (1941), 212-215.

15. M. Filaseta, Sets with elements summing to square-free numbers, C. R. Math. Rep. Acad. Sci. Canada 9 (1987), 243-246.

16. G. Freiman, On two additive problems, to appear.

17. E. Härtter, Ein Beitrag zur Theorie der Minimalbasen, J. reine angew. Math. 196 (1956), 170-204.

18. X.-D. Jia and M. B. Nathanson, A simple construction of minimal asymptotic bases, Acta Arith. 52 (1988), to appear.

19. M. Kneser, Abschätzungen der asymptotischen Dichte von Summenmengen, Math. Zeit. 58 (1953), 459-484.

20. M. B. Nathanson, Minimal bases and maximal nonbases in additive number theory, J. Number Theory 6 (1974), 324-333.

21. M. B. Nathanson, Waring's problem for sets of density zero, in: M. I. Knopp (ed.), Number Theory, Philadelphia 1980, Lecture Notes in Mathematics, Vol. 899, Springer-Verlag, Berlin-New York, 1981, pp. 301-310.

22. M. B. Nathanson, Waring's problem for finite intervals, Proc. Amer. Math. Soc. 96 (1986), 15-17.

23. M. B. Nathanson, A generalization of the Goldbach-Shnirel'man theorem, Amer. Math. Monthly 94 (1987), 768-771.

24. M. B. Nathanson, A short proof of Cauchy's polygonal number theorem, Proc. of the Amer. Math. Soc. 99 (1987), 22-24.

25. M. B. Nathanson, Multiplicative representations of integers, Israel J. Math. 57 (1987), 129-136.

26. M. B. Nathanson, Minimal bases and powers of 2, Acta Arith. 49 (1988), 525-532.

27. M. B. Nathanson, Sumsets containing k-free integers, Journées Arithmétiques de Ulm, 14-18 Septembre 1987, to appear.

28. M. B. Nathanson and A. Sárközy, On the maximum density of minimal asymptotic bases, Proc. Amer. Math. Soc., to appear.

29. M. B. Nathanson and A. Sárközy, Sumsets containing long arithmetic progressions and powers of 2, Acta Arith., to appear.

30. J. Nešetřil and V. Rödl, Two proofs in combinatorial number theory, Proc. Amer. Math. Soc. 93 (1985), 185-188.

31. D. Raikov, Über die Basen der natürlichen Zahlenreihe, Mat. Sbor. N.S. 2 44 (1937), 595-597.

32. L. G. Shnirel'man, Über additive Eigenschaften von Zahlen, Math. Ann. 107 (1933), 649-690.

33. A. Stöhr, Eine Basis h-ter Ordnung für die Menge aller natürlichen Zahlen, Math. Zeit. 42 (1937), 739-743.

34. A. Stöhr, Gelöste und ungelöste Fragen über Basen der natürlichen Zahlenreihe, II, J. reine angew. Math. 194 (1955), 111-140.

35. E. Szemerédi, On sets of integers containing no k elements in arithmetic progression, Acta Arith. 27 (1975), 199-245.

36. E. Wirsing, Thin subbases, Analysis 6 (1986), 285-308.

37. J. Zöllner, Der Vier-Quadrate-Satz und ein Problem von Erdös und Nathanson, Dissertation, Johannes Gutenberg-Universität, Mainz, 1984.

38. J. Zöllner, Über eine Vermutung von Choi, Erdös, und Nathanson, Acta Arith. 45 (1985), 211-213.

# Growth of Order of Homology of Cyclic Branched Covers of Knots

by

Robert Riley

1. In [4] C. McA. Gordon made a study of the groups $H_1\mathfrak{M}_k := H_1(\mathfrak{M}_k,\mathbb{Z})$, $k \geq 2$, where $\mathfrak{M}_k$ is the k-sheeted cyclic cover of $S^3$ branched over a (tame) knot $K \subset S^3$. His Main Theorem was a necessary and sufficient condition for $H_1\mathfrak{M}_k$ to be a periodic function of k. This note concerns an improvement to his preliminary Theorem 4.4, which reads:

> If K is a knot with Alexander polynomial $\Delta(t)$, and some root of $\Delta(t)$ is not a root of unity, then for any integer N there exists k such that $H_1\mathfrak{M}_k$ is finite and order $H_1\mathfrak{M}_k > N$.

Gordon's proof used the standard fact that the order $|H_1\mathfrak{M}_k|$ of $H_1\mathfrak{M}_k$ is the absolute value of the resultant

$$R(k) = R(\Delta(t),t^k-1)$$

$$= c^k \prod_{\ell=1}^{n} (\alpha_\ell^k-1), \text{ when } \Delta(t) = c\prod_{\ell=1}^{n} (\alpha_\ell-t).$$

(By convention, $|H_1\mathfrak{M}_k| = 0$ means $H_1\mathfrak{M}_k$ is an infinite group). He showed that when all roots $\alpha_\ell$ of $\Delta(t)$ of absolute value one are already roots of unity, then the finite values of $|H_1\mathfrak{M}_k|$ grow exponentially with k. However, if $\alpha_\ell$ lies on the unit circle and is not a root of unity, the factor $\alpha_\ell^k - 1$ of R(k) will have arbitrarily small modulus for an infinity of k, and for these k the resultant R(k) could conceivably be nonzero but bounded. Hence the retreat to the irregular growth claimed in Theorem 4.4 above.

Professor W.M. Schmidt of Boulder Colorado pointed out that heavy theorems of A.O. Gel'fond in transcendental number theory preclude the above difficulty. Using Theorem 3.1 of Baker [1], which is a continuation of Gel'fond's work, we shall demonstrate the following improvement to Gordon's Theorem 4.4.

THEOREM. When $K$ is a knot whose Alexander polynomial $\Delta(t)$ has a root which is not a root of unity, then the finite values of $|H_1\mathfrak{M}_k|$ grow exponentially with $k$. More precisely, there are effectively computable constants $a > 0$, $b > 1$, and an effectively computable bound $D$ such that, when $k > D$ then $|H_1\mathfrak{M}_k| > ab^k$.

The effective computability of the constants $a$ and $b$ will be rather obvious, but the effective computability of $D$ is very deep. If we were willing to forgo the effective computability of $D$ the argument could be based on Theorem III of Chapter I of Gel'fond [3], although the proof of this is still quite substantial.

Although we sent a copy of Schmidt's letter to Gordon in 1972 or 73, as far as we know the above Theorem has not become public knowledge. Recent inquiries disclose that the Theorey was still unknown to interested parties, and they would welcome its appearance, hence this note.

2. To prove the Theorem we must establish an exponentially growing lower bound on the absolute values of the nonzero resultants R(k)

defined in §1. We follow Gordon's notation and conventions as closely as possible, thus $\Delta(t) \in \mathbb{Z}[t]$ has $c = \Delta(0) > 0$. Suppose $\Delta(t)$ factors over $\mathbb{Z}[t]$, say $\Delta = \Delta_1 \ldots \Delta_m$ where the factors $\Delta_r(t) \in \mathbb{Z}[t]$ are irreducible, of degree $n_r > 0$. Then

$$R(k) = \prod_{r=1}^{m} R_r(k), \quad R_r(k) = R(\Delta_r(t), t^k - 1).$$

$$\text{If } \Delta_r(t) = c_r \prod_{\ell=1}^{n_r} (\alpha_{\ell,r} - t) \text{ then } R_r(k) = c_r^k \prod_{\ell=1}^{n_r} (\alpha_{\ell,r}^k - 1).$$

When all roots $\alpha_{\ell,r}$ of $\Delta_r$ for some $r$ are roots of unity, then $R_r(k)$ is a periodic function of $k$. The presence of such a periodic factor has no influence on the rate of growth of the nonzero $R(k)$, i.e. on the number $b$ in the theorem, and can only affect the constant multiplier $a$ in an explicitly computable way. It will suffice to show that each irreducible factor $\Delta_r$ whose roots are not roots of unity produces a factor $R_r(k)$ of $|H_1\mathfrak{M}_k|$ which grows exponentially in an effectively computable manner, as specified in the Theorem.

We consider one of these factors $\Delta_r$, and for notational simplicity we omit all reference to the subscript $r$, thus $\Delta_r = \Delta$ has $n_r = n$ roots $\alpha_{\ell,r} = \alpha_\ell$, and highest coefficients $c_r = c$. So

$$R_r(k) = c^k \prod_{\ell=1}^{n} (\alpha_\ell^k - 1).$$

If not all roots lie on the unit circle the product

$$\prod_{\ell, |\alpha_k| < 1} (\alpha_\ell^k - 1) \to 1 \text{ as } k \to \infty.$$

The rate at which this product tends to one will affect $D$, but not $a$ or $b$. The factor

$$S(k) = \prod_{\ell, |\alpha_\ell| > 1} (\alpha_\ell^k - 1)$$

has exponential growth in modulus:

(1) $$|S(k)| > b_1^k,\ b_1 = \prod_{|\alpha_\ell|>1} (|\alpha_\ell| - \varepsilon),$$

when $k >$ computable $D_1(\vec{\alpha}, \varepsilon)$. So if no roots of $\Delta_r$ lie on the unit circle our theorem holds for $\Delta_r$, as noted by Gordon.

The new feature of the analysis concerns the roots on the unit circle. We quote Theorem 3.1 of Baker [1], although we could have used Theorem IV on page 174 of Gel'fond [3].

THEOREM 3.1. Let $\alpha_1, \ldots, \alpha_n$ be non-zero algebraic numbers with degrees at most $d$ and heights at most $A$. Further, let $\beta_0, \ldots, \beta_n$ be algebraic numbers with degrees at most $d$ and heights at most $B(\geq 2)$. Then either

$$\Lambda := \beta_0 + \beta_1 \log \alpha_1 + \ldots + \beta_n \log \alpha_n$$

is zero or $|\Lambda| > b^{-C}$, where $C$ is an effectively computable number depending only on $n$, $d$, $A$, and the original determinations of the logarithms.

The height of an algebraic number $\gamma$ is the maximum of the moduli of the coefficients of a primitive irreducible annilihating polynomial of $\gamma$. In Theorem 2 of [2] Baker gives the explicit estimate $C = (16nd)^{200n}(\log A')^2 \log\log A'$ where $A' = \max(4,A)$, when $\beta_0 = 0$ and $\beta_1, \ldots, \beta_n$ are rational integers.

We apply Theorem 3.1 with Baker's $n = 2$, $\beta_0 = 0$, $\beta_1, \beta_2 \in \mathbb{Z}$, $\alpha_1$ a root of $\Delta(t)$ on the unit circle, and $\alpha_2 = 1$. We take $\log \alpha_1 = ti$ where $0 < t < 2\pi$, and we take $\log 1 = 2\pi i$. Suppose $\alpha_1^k$ is very near one, then $kt/2\pi$ is very near an integer $m(k)$ with $0 < m(k) < k$. Now

$$\Lambda = k \log \alpha_1 - m(k)\cdot 2\pi i$$

cannot vanish because $\alpha_1$ is not a root of unity, hence $|\Lambda| \geq B^{-C}$. Furthermore we have $B = k$ and $C = (32n)^{400}(\log A')^2 \log\log A'$ where $n = \deg \Delta(t)$ again, and $A'$ is immediately derived from $\Delta(t)$. Using $|e^z - 1| > \frac{1}{2}|z|$ for small $|z|$ we deduce

$$|\alpha^k - 1| > \frac{1}{2}k^{-C},$$

valid for all roots of modulus one of $\Delta(t)$. If there are $s$ of these, their combined contribution to the product $|R(k)|$ exceeds $2^{-s}k^{-sC}$.

When all roots of $\Delta(t)$ have modulus one Kronecker's Theorem [5] guarantees that $c \geq 2$, and hence

$$|R(k)| \geq c^k 2^{-n} k^{-nC}.$$

In this case we can take $a = 2^{-n}$, $b = c - \epsilon$ for any $\epsilon$ with $0 < \epsilon < 1$. Then clearly $|R(k)| > a\cdot b^k$, when $k > D(C,c,\epsilon)$, a computable bound. When there are roots off the unit circle we have

$$|R(k)| > c^k \cdot \frac{1}{4}\cdot |S(k)|\cdot 2^{2-n}\cdot k^{(2-n)C} > 2^{-n}(b_1 c)^k k^{(2-n)C},$$

where $k$ is large enough for the contribution from the roots inside $|\alpha| = 1$ to exceed $1/4$, which is a computable bound. It is again clear how to put this in the form $|R(k)| > a\cdot b^k$ when $k > D$, for explicitly computable $a$, $b$, $D$, and our Theorem is proved.

## REFERENCES

[1] A. Baker, "Transcendental Number Theory," Cambridge University Press, Cambridge, 1975.

[2] ________, The Theory of Linear Forms in Logarithms in "Transcendence Theory: Advances and Applications, A Baker and D.W. Masser eds., Academic Press, London, 1977.

[3] A. O. Gel'fond, "Transcendental and Algebraic Numbers," Dover Publications, New York, 1950.

[4] C. McA. Gordon, Knots whose branched cyclic coverings have periodic homology, Trans. Amer. Math. Soc. 168 (1972), 357-370.

[5] L. Kronecker, Zwei Sätze über Gleichungen mit ganzzahlen Coefficienten, J. Reine Angew Math. 53 (1857), 173-175.

Department of Mathematical Sciences
State University of New York at Binghamton
Binghamton, New York 13901

# Hybrid Problems in Number Theory

by

A. Sárközy

Baruch College, The City University of New York,

Department of Mathematics

and

Mathematical Institute of the Hungarian Academy

of Sciences

1. Introduction. The classical number theory studies special sequences of integers (like squares, primes, etc.). It was only about 60-70 years ago that the study of general sequences started by the works of Brun, Šnirel'man and others and it is due mostly to the work of Erdös that the research in this field is getting more and more intensive. ("General sequence" usually means that all we know about the sequence is that it is "dense" in a well-defined sense.) Of course, these two fields of number theory can be combined and one may study problems involving both special sequences and general sequences; a problem of this type can be called "hybrid problem".

The first hybrid problem was studied, perhaps, by Khintchin [28], who proved that the sequence of the squares is an "essential component" (i.e., adding the special sequence of squares to a general sequence of positive Šnirel'man density, we get a sequence of greater Šnirel'man density). Since that, a number of hybrid problems have been studied but the really intensive research in this field started about 10 years ago.

In this paper, first I will give a brief survey of the hybrid problems studied in the last 10 years. Furthermore, in Sections 7 and 8 I will study a special hybrid problem. Finally, in the last section I will list a few related unsolved problems.

2. <u>Notation</u>. $\mathcal{A}, \mathcal{B}, \ldots$ denote sequences of positive integers. The counting function of the sequence $\mathcal{A}$ is denoted by $A(n)$:

$$A(N) = \sum_{\substack{a \le N \\ a \in \mathcal{A}}} 1.$$

$\mathcal{A} + \mathcal{B}$ denotes the set of the integers that can be represented in the form $a + b$ where $a \in \mathcal{A}$, $b \in \mathcal{B}$, and we write $\mathcal{A} + \mathcal{A} = 2\mathcal{A}$, $k\mathcal{A} + \mathcal{A} = (k+1)\mathcal{A}$. $\mathcal{A} - \mathcal{A}$, i.e., the difference set of $\mathcal{A}$ is defined as the set of the positive integers that can be represented in the form $a - a'$ where $a \in \mathcal{A}$, $a' \in \mathcal{A}$. The cardinality of a finite set $S$ is denoted by $|S|$. We write $e^{2\pi i\alpha} = e(\alpha)$. The "norm" of the real number $x$ (the distance from $x$ to the nearest integer) is denoted by $\|x\|$: $\|x\| = \min(\{x\}, 1 - \{x\})$ (where $\{x\}$ denotes the fractional part of $x$). We use the following notation for some arithmetic functions:

$\tau(n)$: the divisor function.

$\sigma(n)$: sum of the divisors of $n$.

$\varphi(n)$: Euler's phi function.

$\mu(n)$: the Moebius function.

$\nu(n)$: the number of distinct prime factors of $n$.

$\omega(n)$: the total number of prime factors of $n$.

$\lambda(n)$: the Liouiville function: $\lambda(n) = (-1)^{\omega(n)}$.

$P(n)$: the greatest prime factor of $n$ $(> 1)$.

$p(n)$: the smallest prime factor of $n$ $(> 1)$.

3. Homogeneous additive hybrid problems. The most general form of an additive hybrid problem is the following: let $\mathcal{A}^{(1)},\mathcal{A}^{(2)},\dots,\mathcal{A}^{(k)}$ be general sequences, let $\mathcal{B}$ be a fixed special sequence, and let $\alpha_1,\dots,\alpha_k$ be real numbers. Then the problem is to study the solvability of the equation

$$\sum_{i=1}^{k} \alpha_i a_{x_i} = b_y, \quad a_{x_i} \in \mathcal{A}^{(1)},\dots,a_{x_k} \in \mathcal{A}^{(k)}, \quad b_y \in B.$$

If $\mathcal{A}^{(1)} = \mathcal{A}^{(2)} = \dots = \mathcal{A}^{(k)}$ and $\alpha_1 + \alpha_2 + \dots + \alpha_k = 0$, then the problem is said to be homogeneous. A typical homogeneous problem is to study the difference set of dense sequences. In fact, the active study of hybrid problems started with the following theorem on difference sets which was proved independently by Furstenberg and me:

If $\mathcal{A}$ is an infinite sequence of positive upper density then

$$a - a' = x^2 \qquad (a,a' \in \mathcal{A},\ x > 0) \tag{1}$$

can be solved.

More exactly, Furstenberg [18] proved the theorem in this form while I [47] proved it in the following quantitative form:

If $N > N_0$ and

$$A(N) \gg N(\log N)^{-1/3}(\log\log N)^{2/3} \tag{2}$$

then (1) can be solved.

Furstenberg used ergodic theory while I adapted that version of the Hardy-Littlewood method which was worked out by Roth in [39] and [40].

In [48] I studied the question that how far is (2) from the best possible. Later Ruzsa [45] improved on this result by showing that there exists a sequence $\mathcal{G}$ such that $\mathcal{G} \subset \{1,2,\ldots,N\}$, $|\mathcal{G}| \gg N^{0.733}$ and (1) cannot be solved.

Erdos asked the question whether the equations

$$(3) \qquad a - a' = p - 1, \quad a,a' \in \mathcal{G}$$

$$(4) \qquad a - a' = x^2 - 1, \quad a,a' \in \mathcal{G}$$

must be solvable in dense sequences $\mathcal{G}$? (Note that the equations $a - a' = p$, $a - a' = x^2 + 1$ need not be solvable in dense sequences as the sequence $\mathcal{G} = \{6,12,18,\ldots\}$ shows.) In [49] I showed that $N > N_0$, $\mathcal{G} \subset \{1,2,\ldots,N\}$, $A(N) \gg N(\log\log N)^{-2}(\log\log\log N)^3(\log\log\log\log N)$ implies the solvability of (3), and by using the same method one could prove an analogous theorem on the solvability of (4). This method is an extension of the method used in [47], and, although not easily, it can be adapted to study any homogeneous additive problem. (See [15] and [44] for estimates from the opposite side.)

Kamae and Mendes France [27] found a third approach to prove the qualitative form of the theorem of Furstenberg and mine. Their approach is based on harmonic analysis. Furthermore, they generalized the problem of solvability of (1) and (4) by studying the more general equation

$$a - a' = f(x)$$

for general polynomials $f(x)$.

Vaughan in his book [60] presented Furstenberg's proof for the theorem on the solvability of (1) wihtout using the terminology of ergodic theory. This presentation suggests that also Furstenberg's method can be modified to get quantitative results, but the estimates obtained in this way would be neither sharper nor simpler.

Very recently Pintz, Steiger and Szemeredi improved on (2) by combining the Hardy-Littlewood method with combinatorial tools (unpublished yet).

Bourgain, Berguelson, Furstenberg and Weiss generalized the problem of solvability of (1) by studying differences of real numbers instead of integers (unpublished yet).

Cantor, Gordon, Erdos, Hartman, Haralambis, Rotenberg, Russa, Stewart and Tijdeman proved several results on difference sets and Ruzsa, Stewart and Tijdeman gave surveys of the results on difference sets [7], [9], [21], [38], [41], [42], [43], [44], [57], [58] and [59].

In [14] and [15] Erdos and I studied both homogeneous and inhomogeneous problems.

4. Inhomogeneous additive problems. Erdos and Turan studied the following problem [17]: if $\mathcal{G}$ is a finite set, then what lower bound can be given in terms of $|\mathcal{G}|$ for the number of distinct prime factors of $\prod_{a,a'\in\mathcal{G}}(a + a')$? Their results have been extended recently by Stewart and Tijdeman.

Lagarias, Odlyzko and Shearer [27], [28] studied the solvability of the equation

$$a + a' = x^2, \quad a \in \mathcal{G}, \quad a' \in \mathcal{G}$$

in "dense" sequences $\mathcal{G}$ (by combining combinatorial and analytical tools).

Starting out from a problem of Erdos, Balog and I showed [3] that if $N > N_0$,

(5) $$\mathcal{A},\mathcal{B} \subset \{1,2,\dots,N\}, \quad |\mathcal{A}|,|\mathcal{B}| \gg N$$

then there exists a "highly composite" sum a + b with

(6) $$P(a+b) < \exp(4(\log N \log\log N)^{1/2}).$$

Furthermore, we proved [4], [5] that if (5) holds then there exists a "near prime" sum a + b with

(7) $$P(a+b) \gg N(\log N)^{-2}$$

and a "near prime square" sum a + b, i.e. a sum a + b and a prime p with

(8) $$p^2 | a+b, \quad p^2 \gg N(\log N)^{-7}.$$

We proved (7) by using the large sieve while in order to prove (6) and (8), we worked out a new version of the Hardy-Littlewood method.

Stewart and I [52] extended problem (7) by studying "near prime" sums of more than two terms, by using Gallagher's larger sieve and the Cauchy-Davenport lemma. In [53] we removed the log power in the denominator of (7), i.e., we showed that (5) implies the existence of a sum a + b with

(9) $$P(a+b) \gg N, \quad a \in \mathcal{A}, \quad b \in \mathcal{B} \quad (\text{for } \mathcal{A}, \mathcal{B} \text{ satisfying (5)})$$

while in [54] we generalized the problem in (8) by studying $k^{th}$ powers instead of squares and we sharpened (8) by removing the log factor. In other words, we proved that for a fixed positive integer $k$ and for $N > N_0(k)$, (5) implies the existence of $a \in \mathcal{A}$, $b \in \mathcal{B}$ and a prime $p$ such that

(10) $$p^k \mid a + b \quad \text{and} \quad p^k \gg N.$$

To prove (9) and (10), we used the same version of the Harchy-Littlewood method which was used by Balog and me in [3], and we added several further ideas. As we realized it later [55], the crucial tool of independent interest in the proof of (9) is an upper bound for $\max_{1/N<\alpha<1-1/N} \sum_{p \le N} \min(y, \|p\alpha\|^{-1})$ (this was only implicit in [52]).

Pomerance, Stewart and I [37] showed that (5) implies the existence of $a \in \mathcal{A}$, $b \in \mathcal{B}$ with

$$p(a+b) = O(1)$$

(by using the large sieve), and we generalized this problem.

Erdos, Pomerance, Stewart and I [13] studied $\max_{a \in \mathcal{A}, b \in \mathcal{B}} \gamma(a+b)$ and, assuming

(11) $$\mathcal{A} \subset \{1,2,\ldots,N\}, \quad |\mathcal{A}| \gg N,$$

the sum $\sum_{a,a' \in \mathcal{A}} \tau(a+a')$ by using combinatorics of finite sets.

Erdos and I studied the following question: how large can $|\mathcal{A}|$ be if $\mathcal{A} \subset \{1,2,\ldots,N\}$ and $a + a'$ is squarefree for all $a \in \mathcal{A}$, $a' \in \mathcal{A}$? The crucial tool in this paper is a "modulo prime square" large sieve.

In [51] I studied the solvability of the equations

$$\lambda(a+b) = +1, \quad a \in \mathcal{A},\ b \in \mathcal{B}$$

$$\lambda(a'+b') = -1, \quad a' \in \mathcal{A},\ b' \in \mathcal{B}$$

for "dense" sequences $\mathcal{A}$, $\mathcal{B}$.

Erdos, Maier and I [10] proved an Erdos-Kac type theorem on sumsets, i.e., we proved that (5) implies that

$$(|\mathcal{A}||\mathcal{B}|)^{-1}|\{(a,b): a \in \mathcal{A},\ b \in \mathcal{B},\ \frac{\nu(a+b) - \log\log N}{(\log\log N)^{1/2}} < x\}|$$

can be approximated by the Gaussian distribution. We used the Hardy-Littlewood method. Elliott and I found another approach, based on a composite moduli large sieve. (Unpublished yet.)

Bourgain, Freiman and Halberstam showed that if $\mathcal{A}$ satisfies (11) then $2\mathcal{A}$ and $3\mathcal{A}$, respectively, contains a "long" arithmetic progression. These results are not exactly hybrid theorems, however, they are theorems of similar nature.

Ostmann [34] raised the following conjecture: There do not exist sequences $\mathcal{A}$ and $\mathcal{B}$ such that they both consist of at least two terms and for $n > n_0$, $n \in (\mathcal{A} + \mathcal{B})$ holds if and only if $n$ is a prime. Hornfeck [24], [25] proved that if $\mathcal{A}$, $\mathcal{B}$ have this property then both of them must be infinite. By using the large sieve, Pomerance, Stewart and I [37] proved that for $\mathcal{A}$, $\mathcal{B}$ with this property we have $A(N)B(N) = O(N)$ but, of course, this is not enough to prove the conjecture. Furthermore we have a lower bound for max $A(N)B(N)$ subject to the condition that $a + b$ is always a prime.

See Erdos and Newman [12], Sarkozy and Szeweredi [56] and Ruzsa [46] for further related results.

5. Partition problems and Ramsey type problems. Erdos and Freud conjectured that

$$1 \le a_1 < a_2 < \ldots < a_{x+1} \le 3x \tag{12}$$

and

$$1 \le b_1 < b_2 < \ldots < b_{x+1} \le 4x \tag{13}$$

imply that there exist a 2-power and a square-free integer that can be written as the sum of distince a's and b's, respectively. This conjecture has been proved by Erdos and Freiman [8] using exponential sums in a way inspired by probability theory. However, their results are not quite satisfactory since they need so many summands as log x, while one might like to show that it suffices to take bounded many summands. Erdos, Nathanson and I [11] showed this (by using Kneser's theorem) in the case when we study the analogous question on infinite sequences. In [33] Nathanson and I showed that also in the finite case. In this paper the proofs are based on the following theorem of independent interest: if $k$ is a positive integer, $\mathcal{G} \subset \{1,2,\dots,N\}$ and $|\mathcal{G}| > \frac{N}{k} + 1$, then there exists a positive integer $d$ such that $d \leq k - 1$ and $(4k)\mathcal{G}$ contains an arithmetic progression of $[\frac{N}{2kd}]$ terms and of difference $d$.

Burr, Erdos and I discussed the following questions: is it true that if $k$ is a positive integer and we split the sequence of squares into k classes, then for $n > n_0$, $n$ can be represented as the sum of squares belonging to the same class (in other words, as a "monochromatic"sum of squares)? Can $n$ be represented by distinct squares taken from the same class? Nothing has been published yet.

See Hindman [23] for a further related result.

6. Multiplicative hybrid problems. I proved [50] that if $p$ is a prime number, $\mathcal{G}, \mathcal{B} \subset \{1,2,\dots,p-1\}$, $|\mathcal{G}| \gg p$ and $|\mathcal{B}| \gg p$, then the least non-negative residues of the products ab ($a \in \mathcal{G}$, $b \in \mathcal{B}$) are uniformly distributed with an error term $c\sqrt{p}\log p$ (in other words, a Polya-Vinogradov type inequality holds). This implies that there exists a product whose least non-negative residue is $\ll \sqrt{p}\log p$.

This problem generalizes the problem of the least quadratic non-residue modulo $p$, in fact, if $\mathcal{A}$ is the set of the quadratic residues while $\mathcal{B}$ is the set of the quadratic non-residues modulo $p$, then the products ab are the quadratic non-residues modulo $p$. Balog has shown recently that also the Burgess inequality can be adapted to this situation (unpublished yet).

Irvaniec and I [26] proved the following theorem (by using exponential sums): if $\mathcal{A}$, $\mathcal{B}$ satisfy (5) then there exists a product ab (where $a \in \mathcal{A}$, $b \in \mathcal{B}$) and a square $x^2$ such that

$$|ab - x^2| \ll (N \log N)^{1/2}.$$

Pomerance and I [35] studied "homogeneous" multiplicative problems (by using combinatorial tools) and we sharpened a classical theorem of Behrend [6] by showing that if $\mathcal{A} \subset \{1,2,\ldots,N\}$, $\mathcal{P}$ is a set of prime numbers not exceeding $N$, and both $\mathcal{A}$ and $\mathcal{P}$ are "dense" in a logarithmic density sense, i.e., $\Sigma\frac{1}{a}$ and $\Sigma\frac{1}{p}$ are "large", then there exist $a \in \mathcal{A}$, $a' \in \mathcal{A}$ such that $a|a'$ and all the prime factors of $\frac{a'}{a}$ belong to $\mathcal{P}$. On the other hand, we showed that, e.g., $a|a'$, $\frac{a'}{a} = p - 1$ need not be solvable (it is trivial that $\frac{a'}{a} = p$ need not be solvable).

Pomerance and I [36] showed that if $\mathcal{A}$ and $\mathcal{B}$ satisfy (5) then the number of the distinct products of the form ab (where $a \in \mathcal{A}$, $b \in \mathcal{B}$) is greater, than $N^2(\log N)^{1-2\log 2+\varepsilon}$, and we used this theorem to show the existence of integers a,b,a',b',n and a prime $p$ such that $a \in \mathcal{A}$, $b \in \mathcal{B}$, $P(n)$ is "small" and both $|ab-p|$ and $|a'b'-n|$ are small (e.g., if the Riemann hypothesis is true then $\min|ab-p| < (\log N)^{1+2\log 2+\varepsilon}$).

7. <u>An inhomogeneous additive problem.</u> Our results (9) and (10) with Stewart inspired the following conjecture: if $\mathcal{A}$, $\mathcal{B}$ satisfy (5), then there exist integers a, b with

$$P(a^2+b^2) \gg N^2, \quad a \in \mathcal{A},\ b \in \mathcal{B}. \tag{14}$$

This problem seems to be very difficult (if not hopeless). In the next two sections I will discuss those weaker results that I have been able to prove in connection with this problem.

THEOREM 1. If $N > N_0$, $\mathcal{A} \subset \{1,2,\dots,N\}$, $\mathcal{B} \subset \{1,2,\dots,N\}$ and

$$|\mathcal{A}||\mathcal{B}| > 225\, N(\log N)^2 \tag{15}$$

then there exist integers a, b such that $a \in \mathcal{A}$, $b \in \mathcal{B}$ and

$$P(a^2+b^2) > \frac{1}{30}\,\frac{(|\mathcal{A}||\mathcal{B}|)^{1/2}}{\log N}.$$

(So that $|\mathcal{A}| \gg N$, $|\mathcal{B}| \gg N$ implies $\max_{a\in\mathcal{A},b\in\mathcal{B}} P(a^2+b^2) \gg \frac{N}{\log N}$.)

PROOF. The proof will be based on the following large sieve result.

LEMMA 1. Let $\eta$ be a set of $Z$ integers in the interval $[M+1,M+N]$. For proime $p$ let $\omega(p)$ denote the number of residue classes modulo p that contain no element of $\eta$. Then for $Q \geq 1$ we have

$$Z \leq \frac{N+2Q^2}{L}$$

where

$$L = \sum_{q\leq Q} \mu^2 q \prod_{p|q} \frac{\omega(p)}{p-\omega(p)}.$$

See [31], p. 25.

In order to prove Theorem 1, we put

$$Q = \frac{1}{15} \frac{(|\mathcal{A}||\mathcal{B}|)^{1/2}}{\log N},$$

and we start out from the indirect assumption

$$P(a^2+b^2) \le Q/2 \tag{16}$$

(for all $a \in \mathcal{A}$, $b \in \mathcal{B}$) which implies that

$$p \nmid (a^2+b^2) \quad \text{for } Q/2 < p \le Q,\ a \in \mathcal{A},\ b \in \mathcal{B}. \tag{17}$$

For $Q/2 < p \le Q$, let $\varphi(p)$ and $\psi(p)$ denote the number of residue classes modulo $p$ that contain no element of $\mathcal{A}$ and $\mathcal{B}$, respectively. Let $p \equiv 1 \pmod 4$. If $\mathcal{A}$ contains an element of the $0$ residue class modulo $p$, then by (17), $\mathcal{B}$ must not intersect the $0$ residue class. Furthermore, if $(r,p) = 1$ and $\mathcal{A}$ contains an element of one of the residue classes represented by the integers $r$ and $-r$, respectively, and $s$ is an integer with

$$r^2 + s^2 \equiv 0 \pmod p$$

(such an integer $s$ exists by $p \equiv 1 \pmod 4$, and clearly, $s \not\equiv 0 \pmod p$)), then by (17), $\mathcal{B}$ must not intersect the residue classes represented by $s$ and $-s$, respectively. (Clearly, this is a one-to-one correspondence between the pairs of residue classes $(r,-r)$ and $(s,-s)$, respectively.) It follows from this discussion that

$$\varphi(p) + \psi(p) \ge p \qquad (\text{for } p \equiv 1 \pmod 4)). \tag{18}$$

By using Lemma 1 with $\mathcal{A}$ and $\mathcal{B}$, respectively, in place of $\mathcal{R}$, we obtain that

$$(19)\qquad |\mathfrak{C}| \le (N+2Q^2)\Big(\sum_{q\le Q} \mu^2(q) \prod_{p|q} \frac{\varphi(p)}{p-\varphi(p)}\Big)^{-1}$$

$$\le (N+2Q^2)\Big(\sum_{\substack{Q/2<p\le Q\\ p\equiv 1 \pmod 4}} \mu^2(p)\,\frac{\varphi(p)}{p-\varphi(p)}\Big)^{-1}$$

$$= (N+2Q^2)\Big(\sum_{\substack{Q/2<p\le Q\\ p\equiv 1 \pmod 4}} \frac{\varphi(p)}{p-\varphi(p)}\Big)^{-1},$$

and, in view of (18),

$$(20)\qquad |\mathfrak{D}| \le (N+2Q^2)\Big(\sum_{q\le Q} \mu^2(q) \prod_{p|q} \frac{\psi(p)}{p-\psi(p)}\Big)^{-1}$$

$$\le (N+2Q^2)\Big(\sum_{\substack{Q/2<p\le Q\\ p\equiv 1 \pmod 4}} \mu^2(p)\,\frac{\psi(p)}{p-\psi(p)}\Big)^{-1}$$

$$\le (N+2Q^2)\Big(\sum_{\substack{Q/2<p\le Q\\ p\equiv 1 \pmod 4}} \frac{p-\varphi(p)}{\varphi(p)}\Big)^{-1}$$

since $0 < u,v < 1$ and $u + v \ge 1$ imply that $\frac{u}{1-u} \ge \frac{1-v}{v}$. (Note that (17) implies $p > \varphi(p) > 0$, $p > \psi(p) > 0$.) By using Cauchy's inequality and in view of (15), we obtain from (19) and (20) that for large $N$ (so that also $Q$ is large by (15)) we have

$$|\mathfrak{C}||\mathfrak{D}| \le (N+2Q^2)^2\Big(\sum_{\substack{Q/2<p\le Q\\ p\equiv 1 \pmod 4}} \frac{\varphi(p)}{p-\varphi(p)} \sum_{\substack{Q/2<p\le Q\\ p\equiv 1 \pmod 4}} \frac{p-\varphi(p)}{\varphi(p)}\Big)^{-1}$$

$$\le (N+2Q^2)^2\Big(\sum_{\substack{Q/2<p\le Q\\ p\equiv 1 \pmod 4}} \Big(\frac{\varphi(p)}{p-\varphi(p)}\Big)^{1/2}\Big(\frac{p-\varphi(p)}{\varphi(p)}\Big)^{1/2}\Big)^{-2}$$

$$= (N+2Q^2)^2\Big(\sum_{\substack{Q/2<p\le Q\\ p\equiv 1 \pmod 4}} 1\Big)^{-2} < (N+2Q^2)^2\Big(\frac{1}{5}\,\frac{Q}{\log Q}\Big)^{-2}$$

$$< 25Q^2\Big(\frac{N}{Q^2}+2\Big)^2(\log Q)^2 =$$

$$= 25\cdot\frac{1}{225}\,\frac{|\mathcal{A}||\mathcal{B}|}{(\log N)^2}\left(\frac{225\,N(\log N)^2}{|\mathcal{A}||\mathcal{B}|}+2\right)^2\left(\log\left(\frac{1}{15}\,\frac{(|\mathcal{A}||\mathcal{B}|)^{1/2}}{\log N}\right)\right)^2$$

$$< \frac{1}{9}\,\frac{|\mathcal{A}||\mathcal{B}|}{(\log N)^2}\cdot 3^2(\log N)^2 = |\mathcal{A}||\mathcal{B}|,$$

and this contradiction shows that the indirect assumption (16) cannot hold which completes the proof of Theorem 1.

8. Another approach to the same problem. Another approach to the problem discussed in Section 7 is to prove the analogue of (14) with more squares in place of $a^2 + b^2$.

THEOREM 2. For all $\varepsilon > 0$ there exist numbers $N_0 = N_0(\varepsilon)$ and $c = c(\varepsilon)$ $(> 0)$ such that if $N > N_0$,

(21) $$\mathcal{A} \subset \{1,2,\dots,N\} \quad \text{and} \quad |\mathcal{A}| > \varepsilon N,$$

then there exist integers $a_1,\dots,a_8$ with $a_1 \in \mathcal{A},\dots,a_8 \in \mathcal{A}$ and

(22) $$P(a_1^2 +\dots+ a_8^2) > c(\varepsilon)N^2.$$

PROOF. Let $\mathcal{B}$ denote the set of the integers $n$ that can be represented in the form

(23) $$a_1^2 + a_2^2 + a_3^2 + a_4^2 = n, \quad a_1 \in \mathcal{A},\ a_2 \in \mathcal{A},\ a_3 \in \mathcal{A},\ a_4 \in \mathcal{A}.$$

First we are going to give a lower bound for $|\mathcal{B}|$.

Let $f(n)$ and $g(n)$ denote the number of solutions of (23) and

$$x^2 + y^2 + z^2 + u^2 = n$$

(where $x,y,z,u$ are integers), respectively. By the Cauchy-Schwarz inequality we have

(24) $$|\mathfrak{B}| = \sum_{n\in\mathfrak{B}} 1 \geq \Big(\sum_{n\in\mathfrak{B}} f(n)\Big)^2 \Big(\sum_{n\in\mathfrak{B}} f^2(n)\Big)^{-1}$$

$$\geq \Big(\sum_{n\in\mathfrak{B}} f(n)\Big)^2 \Big(\sum_{n=1}^{4N^2} g^2(n)\Big)^{-1}.$$

Clearly we have

(25) $$\sum_{n\in\mathfrak{B}} f(n) = \sum_{a_1,a_2,a_3,a_4\in\mathfrak{a}} = |t|^4 .$$

Thus in order to give a lower bound for $|\mathfrak{B}|$, it suffices to give an upper bound for $\sum_{n=1}^{4N^2} g^2(n)$. It is well-known (see, e.g., [22], p.3.4) that

$$(g(n) = 8 \sum_{\substack{d|n \\ 4\nmid d}} d \leq 8\sigma(n)$$

hence

(26) $$\sum_{n=1}^{4N^2} g^2(n) \leq 64 \sum_{n=1}^{4N^2} \sigma^2(n).$$

LEMMA 2.

$$\sum_{n=1}^{N} \sigma^2(n) = \Big(\frac{C}{3} + o(1)\Big)N^3$$

where

$$C = \prod_p \Big(1 + \frac{2p^3+p^2-1}{(p^2-1)(p^3-1)}\Big).$$

(Clearly, this infinite product is convergent.)

PROOF. Define the function $h(n)$ by

(27) $$\Big(\frac{\sigma(n)}{n}\Big)^2 = \sum_{d|n} h(d).$$

Then by the Moebius inversion formula $h(d)$ is multiplicative and

$$h(n) = \sum_{d|n} \mu(d) \left(\frac{\sigma(n/d)}{n/d}\right)^2$$

hence

(28)
$$h(p^\alpha) = \left(\frac{\sigma(p^\alpha)}{p^\alpha}\right)^2 - \left(\frac{\sigma(p^{\alpha-1})}{p^{\alpha-1}}\right)^2$$

$$= \left(\frac{p^{\alpha+1}-1}{p^\alpha(p-1)}\right)^2 - \left(\frac{p^\alpha-1}{p^{\alpha-1}(p-1)}\right)^2 = \frac{2p^{\alpha+1}-p-1}{p^{2\alpha}(p-1)}.$$

This implies that

(29)
$$0 < h(p^\alpha) < \frac{2p^{\alpha+1}}{p^{2\alpha}(p/2)} = \frac{4}{p^\alpha}$$

hence

(30)
$$0 < h(n) \le \frac{4^{\nu(n)}}{n} < n^{-1} \exp\left(c \frac{\log n}{\log\log n}\right).$$

(30) implies that

(31)
$$\sum_{n=1}^{N} h(n) = \sigma(N)$$

and

(32)
$$\sum_{n=1}^{+\infty} \frac{h(n)}{n} < +\infty.$$

In view of (27), (29), (30), (31) and (32) we have

(33)
$$\sum_{n=1}^{N} \left(\frac{\sigma(n)}{n}\right)^2 = \sum_{n=1}^{N} \sum_{d|n} h(d) = \sum_{d=1}^{N} h(d)\left[\frac{N}{d}\right]$$

$$= N \sum_{d=1}^{N} \frac{h(d)}{d} + O\left(\sum_{d=1}^{N} h(d)\right) = N \sum_{d=1}^{+\infty} \frac{h(d)}{d} + \sigma(N)$$

$$= N \prod_{p} \left(1 + \sum_{\alpha=1}^{+\infty} \frac{h(p^\alpha)}{p^\alpha}\right) + \sigma(N).$$

By (26) here we have

$$(34)\qquad \sum_{\alpha=1}^{+\infty} \frac{h(p^\alpha)}{p^\alpha} = \sum_{\alpha=1}^{+\infty} \frac{2p^{\alpha+1}-p-1}{p^{3\alpha}(p-1)}$$

$$= \frac{2p}{p-1} \sum_{\alpha=1}^{+\infty} p^{-2\alpha} - \frac{p+1}{p-1} \sum_{\alpha=1}^{+\infty} p^{-3\alpha}$$

$$= \frac{2p}{(p-1)(p^2-1)} - \frac{p+1}{(p-1)(p^3-1)} = \frac{2p^3+p^2-1}{(p^2-1)(p^3-1)}.$$

(33) and (34) imply that

$$\sum_{n=1}^{N} \left(\frac{\sigma(n)}{n}\right)^2 = (C + o(1))N$$

and the statement of the lemma follows by partial summation.

It follows from (24), (25), (26) and Lemma 2 (with $4N^2$ in place of $N$) that

$$|\mathfrak{B}| \gg |\mathcal{Q}|^8 \Big( \sum_{n=1}^{4N^2} \sigma^2(n) \Big)^{-1} \gg |\mathcal{Q}|^8 N^{-6} > \varepsilon^8 N^2 .$$

Thus writing $M = 4N^2$, we have $\mathfrak{B} = \{1,2,\ldots,M\}$ and

$$|\mathfrak{B}| \gg \varepsilon^8 M^2 .$$

By a theorem of Stewart and mine [52] (see (9) above) this implies that for $N > N_1(\varepsilon)$ there exist $b \in \mathfrak{B}$, $b' \in \mathfrak{B}$ with

$$(35)\qquad P(b+b') > c(\varepsilon)M > c'(\varepsilon)N^2 .$$

By $b \in \mathfrak{B}$, $b' \in \mathfrak{B}$, both $b$ and $b'$ can be represented in the form $a_1^2 + a_2^2 + a_3^2 + a_4^2$ (where $a_1,a_2,a_3,a_4 \in \mathcal{Q}$) and thus $b + b'$ can be represented in the form $a_1^2 + a_2^2 +\ldots+ a_8^2$. In view of (35), this sum satisfies (22) and this completes the proof of Theorem 2.

9. Unsolved problems. Finally, I am going to list some related unsolved problems (including some non-hybrid, however related questions).

Problem 1. Improve on Theorems 1 and 2 by showing that if $\varepsilon > 0$, $N > N_0(\varepsilon)$, $\mathcal{A} \subset \{1,2,\dots,N\}$, $\mathcal{B} \subset \{1,2,\dots,N\}$, $|\mathcal{A}| \gg N$, $|\mathcal{B}| \gg N$ then there exist integers $a, b$ with $a \in \mathcal{A}$, $b \in \mathcal{B}$, $P(a^2+b^2) \gg N^{2-\varepsilon}$ (and, once more, I conjecture that this holds even with $N^2$ in place of $N^{2-\varepsilon}$).

Problem 2. Our papers [1] and [2] with Balog lead me to the following conjecture: if $\varepsilon > 0$, $n > n_0(\varepsilon)$, then

$$(36) \qquad x^2 + y^2 + z^2 + u^2 = n, \quad P(x) < n^{\varepsilon},\ P(y) < n^{\varepsilon},\ P(z) < n^{\varepsilon},\ P(u) < n^{\varepsilon}$$

can be solved. This seems to be difficult but, perhaps, as a first step it can be shown that (36) is solvable for almost all $n$.

Problem 3. Another problem inspired by the same papers: is it true that for $\varepsilon > 0$, $n > n_0(\varepsilon)$,

$$x^2 + y^2 + z = n, \quad P(z) < n^{\varepsilon}$$

can be solved? (Perhaps, Linnik's dispersion method can be used.)

Problem 4. Perhaps, the problem studied in [1] and [2] can be generalized in the following way: if $\varepsilon > 0$, $n > n_0(\varepsilon)$, $\mathcal{P}$ is a set of prime numbers not exceeding $n$ with $2 \in \mathcal{P}$ and

$$\sum_{p \in \mathcal{P}} \frac{1}{p} > \left(\frac{1}{2}+\varepsilon\right) \log\log n,$$

then $n$ can be represented as the sum of two integers all whose prime factors belong to $\mathcal{P}$. It seems to be hopeless to prove this, but, perhaps, it can be shown that $n$ can be represented as the sum of <u>three</u> integers of this form.

Problem 5. Show that if $f(n)$ is a multiplicative arithmetic function such that $f(p^\alpha) = -1$ or $f(p^\alpha) = +1$ for all prime powers $p^\alpha$, and

$$\sum_{p \equiv a \pmod q} \frac{1}{p} = +\infty \quad \text{for} \quad q = 1,2,\ldots \text{ and } (a,q) = 1,$$

then we have

$$\max_{0 \le \alpha \le 1} \left| \sum_{n=1}^{N} f(n)e(nx) \right| = o(N).$$

By using this, one could generalize the theorem in [51] and some results in [14]. For $\alpha$'s belonging to the "minor arcs", one can estimate the sum in (37) by using a quantitative version of Daboussi's theorem given by Montgomery and Vaughan [32], while for $\alpha$'s belonging to the "major arcs", one might like to use the mean value theorem of Halasz [19], [20] with a good error term; unfortunately, such an estimate is not available.

Problem 6. Hardy and Ramanujan showed that almost all the integers $n$ whose decimal form is $9\ldots9$ have less than $(1+\varepsilon)\log\log n$ prime factors. (In fact, they proved a slightly more general statement.) Partly this theorem, partly our paper with Erdos and Maier [10] inspired the following question: is it true, that almost all the integers $n$, whose decimal form does not contain, say, the digit 9, have about $\log\log n$ prime factors?

Problem 7. Improve on the results in [11] by determining the smallest integers $k$ and $\ell$ with the property that if $\mathcal{A}, \mathcal{B}$ are infinite sequences of integers with

$$A(3x) = \sum_{\substack{a \leq 3x \\ a \in \mathcal{A}}} 1 \geq x + 1 \qquad (\text{for } x > x_0)$$

and

$$B(4x) = \sum_{\substack{b \leq 4x \\ b \in \mathcal{B}}} 1 \geq x + 1 \qquad (\text{for } x > x_1),$$

then there exist a 2-power and a square-free integer that can be represented as the sum of at most $k$ and $\ell$ terms of the sequences $\mathcal{A}$ and $\mathcal{B}$, respectively. (Erdos, Nathanson and I [11] showed that $k = 5$ and $\ell = 6$ can be chosen. Perhaps, $\min k = 3$ and $\min \ell = 2$. The analogous question on finite sequences seems to be much more difficult.)

Problem 8. Generalize the result in [51] by showing that if $k$ is a positive integer and $\mathcal{A}$, $\mathcal{B}$ satisfy (5), then there exist integers $a, b$ with $a \in \mathcal{A}$, $b \in \mathcal{B}$ and $k | \nu(a+b)$.

Problem 9. Pomerance and I [35] conjectured that Behrend's theorem [6] can be sharpened in the following way: if $k$ is a fixed integer, $N$ is an integer with $N > N_0(k)$, $\mathcal{A} \subset \{1,2,\ldots,N\}$ and

$$\sum_{a \in \mathcal{A}} \frac{1}{a} > c \frac{\log N}{(\log \log N)^{1/2}})$$

then there exist integers $a$, $a'$ ($\neq a$) such that $a \in \mathcal{A}$, $a' \in \mathcal{A}$, $a | a'$ and $\frac{a'}{a} \equiv 1 \pmod{k}$.

This assertion would follow from the following one: if $k$ is a fixed integer, $N$ is an integer with $N > N_1(k)$, $\mathcal{A} \subset \{1,2,\ldots,N\}$, $a \equiv a' \pmod{k}$ for all $a \in \mathcal{A}$, $a' \in \mathcal{A}$, and

$$\sum_{a \in \mathcal{G}} \frac{1}{a} > c' \frac{\log N}{k(\log \log N)^{1/2}},$$

then there exist integers a, a' ($\neq$ a) such that a $\in \mathcal{G}$, a' $\in \mathcal{G}$ and $a | a'$.

REFERENCES


[1] A. Balog and A. Sarkozy, On sums of integers having small prime factors, I, Studia Sci. Math. Hung. 19 (1984), 35-47.

[2] ______________________, II, Studia Sci. Math. Hung. 19 (1984), 81-89.

[3] ______________________, On sums of sequences of integers, I, Acta Arithmetica 44 (1984), 73-86.

[4] ______________________, II, Acta Math. Acad. Sci. Hung. 44 (1984), 169-179.

[5] ______________________, III, Acta Math. Acad. Sci. Hung. 44 (1984), 339-349.

[6] F. Behrend, On sequences of numbers not divisible one by another, J. London Math. Soc. 10 (1935), 42-45.

[7] D. G. Cantor and B. Gordon, Sequences of integers with missing differences, J. Comb. Th. Ser. A 14 (1973), 281-287.

[8] P. Erdos and G. Freiman, On two additive problems, to appear.

[9] P. Erdos and S. Hartman, On sequences of distances of a sequence, Colloq. Math. 17 (1967), 191-193.

[10] P. Erdos, H. Maier and A. Sarkozy, On the distribution of the number of prime factors of sums a + b, Transactions Amer. Math. Soc., to appear.

[11] P. Erdos, M.B. Nathanson and A. Sarkozy, Sumsets containing infinite arithmetic progressions, J. Number Theory, to appear.

[12] P. Erdos and D.J. Newman, Bases for sets of integers, J. Number Th. 9 (1977), 420-425.

[13] P. Erdos, C. Pomerance, A. Sarkozy and C.L. Stewart, On extreme values of arithmetic functions on sums, to appear.

[14] P. Erdos and A. Sarkozy, On differences and sums of integers, I, J. Number Theory 10 (1978), 430-450.

[15] __________________, II, Bull. Soc. Math. Grece 18 (1977), 204-223.

[16] __________________, On divisibility properties of integers of the form a + a', Acta Math. Acad. Sci. Hung., to appear.

[17] P. Erdos and P. Turan, On a problem in the elementary theory of numbers, Amer. Math. Monthly 41 (1934), 608-611.

[18] H. Furstenberg, Ergodic behavior of diagonal measures and a theorem of Szemeredi on arithmetic progressions, J. Analyse Math. 31 (1977), 204-256.

[19] G. Halasz,Uber die Mittelwerte multiplikativer zahlentheoretischer Functionen, Acta Math. Acad. Sci. Hung. 19 (1968), 365-403.

[20] G. Halasz, On the distribution of additive and the mean values of multiplicative arithmetic functions, Studia Sci. Math. Hung. 6 (1971), 211-233.

[21] N.M. Haralambis, Sets of integers with missing differences, J. Comb. Th. Ser. A 23 (1977), 22-33.

[22] G. H. Hardy and E.M. Wright, An Introduction to the Theory of Numbers, 5th edition, Oxford University Press, 1979.

[23] N. Hindman, Finite sums from sequences within cells of a partition of N, J. Comb. Th. Ser. A 17 (1974), 1-11.

[24] B. Hornfeck, Ein Satz uber die Primzahlmenge, Math. Z. 60 (1954), 271-273.

[25] B. Hornfeck, Berichtigung zur Arbeit: Ein Satz uber die Primzahlmenge, Math. Z. 62 (1955), 502.

[26] H. Iwaniec and A. Sarkozy, On a multiplicative hybrid problem, J. Number Theory, to appear.

[27] T. Kamae and M. Mendes France, Van der Corput's difference theorem, Israel J. Math. 31 (1978), 335-342.

[28] A. Khintchin, Zur additiven Zahlentheorie, Mat. Sb. N.S. 39 (1932), 27-34.

[29] J.C. Lagarias, A. M. Odlyzko and J.B. Shearer, On the density of sequences of integers the sum of no two of which is a square, I, J. Combinatorial Theory 33 (1982), 167-185.

[30] ____________________________, II, J. Combinatorial Theory 34 (1983), 123-139.

[31] H. L. Montgomery, Topics in Multiplicative Number Theory, Lecture Notes in Mathematics, 227, Springer-Verlag, 1971.

[32] H. L. Montgomery and R.C. Vaughan, Exponential sums with multiplicative coefficients, Invent. Math. 43 (1977), 69-82.

[33] M. B. Nathanson and A. Sarkozy, Sumsets containing long arithmetic progressions, 2-powers and square-free integers, to appear.

[34] H. Ostmann, Additive Zahlentheorie, Springer-Verlag, 1956.

[35] C. Pomerance and A. Sarkozy, On homogeneous multiplicative hybrid problems in number theory, Acta Arithmetica, to appear.

[36] C. Pomerance and A. Sarkozy, On products of sequences of integers, to appear.

[37] C. Pomerance, A. Sarkozy and C.L. Stewart, On divisors of sums of integers, III, Pacific Journal, to appear.

[38] D. Rotenberg, Sur une classe de parties evitables, Colloq. Math. 20 (1969), 67-68.

[39] K. F. Roth, Sur quelques ensembles d'entriers, C.R. Acad. Sci. Paris 234 (1952), 388-390.

[40] K. F. Roth, On certain sets of integers, J. London Math. Soc. 28 (1953), 104-109.

[41] I. Z. Ruzsa, On difference-sequences, Acta Arith. 25 (1973/74), 151-157.

[42] I. Z. Ruzsa, On difference sets, Studici Sci. Math. Hungar. 13 (1978), 319-326.

[43] I.Z. Ruzsa, Uniform distribution, positive trigonometric polynomials and difference sets, Seminar on Number Theory, 1981-82, Exp. No. 18, 78 pp., Univ. Bordeaux Talence, 1982.

[44] I.Z. Ruzsa, On measures on intersectivity, Acta Math. Hungar. 43 (1984), 335-340.

[45] I.Z. Ruzsa, Difference sets without squares, Period. Math. Hungar. 15 (1984), 205-209.

[46] I.Z. Ruzsa, On an additive property of squares and primes, to appear.

[47] A. Sarkozy, On difference sets of sequences of integers, I, Acta Math. Acad. Sci. Hung. 31 (1978), 125-149.

[48] __________, II, Annales Univ. Sci. Budapest, Eotros 21 (1978), 45-53.

[49] __________, III, Acta Math. Acad. Sci. Hung. 31 (1978), 355-386.

[50] __________, On the distribution of residues of products of integers, Acta Math. Acad. Sci. Hung., to appear.

[51] __________, On the number of prime factors of integers of the form $a_i + b_j$, Studia Sci. Math. Hung., to appear.

[52] A. Sarkozy and C.L. Stewart, On divisors of sums of integers, I, Acta Math. Acad. Sci. Hung. 48 (1986), 147-154.

[53] ____________________, II, J. Reine Angew. Math. 365 (1986), 171-191.

[54] ____________________, IV, Canadian J., to appear.

[55] ____________________, On sums of the form $\Sigma_p \min(y, \|p\alpha\|^{-1})$, J. Australian Math. Soc., to appear.

[56] A. Sarkozy and E. Szemeredi, On the sequence of squares (in Hungarian), Mat. Lapok 16 (1965), 76-85.

[57] C. L. Stewart, On difference sets of sets of integers, Seminaire Delange-Pisot-Poitou (Theorie des numbres). 19 (5) (1977-78), 8 p.

[58] C. L. Stewart and R. Tijdeman, On infinite difference sets, Canadian J. Math. 31 (1979), 897-910.

[59] R. Tijdeman, Distance sets of sequences of integers, Proceedings, Bicentennial Congress Wiskundig Genootschap (Vrije Univ., Amsterdam, 1978), Part II, pp. 405-415, Math. Centre Tracts, 101, Math. Centrum, Amsterdam, 1979.

[60] R. C. Vaughan, The Hardy-Littlewood Method, Cambridge, 1981.

# Binomial Coefficients Not Divisible by a Prime

Alan H. Stein

## 1 Introduction

Questions relating to binomial coefficient parity have been studied by several researchers, including Fine, Harborth, Singmaster, Stolarsky and the author. It is known in a fairly precise way that almost all binomial coefficients are even [4,13].

In this paper, we shall exhibit analogous results for all primes.

Let $p$ represent a fixed prime $p$, $T(n) = T_p(n)$ be the number of binomial coefficients $\binom{n}{m}$ not divisible by $p$ and let $\phi(x) = \phi_p(x) = \sum_{n<x} T(n)$. We can interpret $\phi(x)$ as the number of binomial coefficients in the first $x$ rows of Pascal's triangle which are not divisible by $p$.

Fine [2] showed that $\phi(x) = o(x^2)$ and thus most binomial coefficients are divisible by $p$. We will obtain asymptotic bounds on $\phi(x)$ similar to those previously obtained for the binary case and use these bounds to show that, given a finite set of primes, almost all binomial coefficients are divisible by all primes in that set.

## 2 Basic Properties and Inequalities

Notation

$$p = \text{a fixed prime} \quad (1)$$

$$P = p(p+1)/2 \quad (2)$$

$$\theta = \theta_p = \log_p P \quad (3)$$

$$T(n) = T_p(n) = \text{the number of binomial coefficients } \binom{n}{m} \text{ not divisible by p} \quad (4)$$

$$\phi(x) = \phi_p(x) = \sum_{n<x} T(n) \quad (5)$$

$$\psi(x) = \psi_p(x) = \phi(x)/x^\theta \quad (6)$$

For any integer $n$, we will let $n_i$ represent the digit in the $p^i th$ place in its $p$-ary representation, i.e. $n = \sum n_i p^i$. Fine [2] and Singmaster [6,7,8], among others, have shown that

$$T(n) = \prod (n_i + 1). \quad (7)$$

This is an immediate consequence of the well known result that $\binom{n}{m}$ is divisible by $p$ if and only $n_i < m_i$ for some $i$. In the special case $n < p$, this reduces to $T(n) = n + 1$, so

$$\phi(x) = \sum_{n<x} (n+1) = x(x+1)/2 \text{ for } x \leq p. \tag{8}$$

As a special case,

$$\phi(p) = P,\ \psi(p) = 1$$

and Fine has shown that

$$\phi(p^k) = P^k,\ \psi(p^k) = 1.$$

From 7, it immediately follows that

$$T(mp^k + n) = T(m)T(n) \text{ for } n < p^k. \tag{9}$$

**Lemma 1** $\phi(px) = P\phi(x)$.

**Proof:**

$$\begin{aligned}\phi(px) &= \sum_{m<p,n<x} T(np+m) = \sum_{m<p,n<x} T(m)T(n) \\ &= \phi(p)\phi(x) = P\phi(x).\end{aligned}$$

∎

**Corollary 2** $\psi(px) = \psi(x)$.

This shows that the periodicity of $\psi$ observed for $p = 2$ extends to all primes $p$.

**Lemma 3** $\phi(p^k x + y) = P^k\phi(x) + T(x)\phi(y)$ *for* $y \leq p^k$.

**Proof:**

$$\begin{aligned}\phi(p^k x + y) &= \phi(p^k x) + \sum_{n<y} T(p^k x + n) \\ &= P^k\phi(x) + \sum_{n<y} T(x)T(n) = P^k\phi(x) + T(x)\phi(y).\end{aligned}$$

∎

Using Lemma 3, we easily show that $\psi$ is bounded above and below.

**Lemma 4** $1/P \leq \psi(x) \leq P$.

**Proof:** Let $p^k \leq x \leq p^{k+1}$. Then $1/P = \phi(p^k)/(p^{k+1})^\theta \leq \phi(x)/x^\theta = \psi(x) \leq \phi(p^{k+1})/(p^k)^\theta = P$.

∎

As is already known for the case $p = 2$, the upper bound for $\psi$ can be improved to 1 and $\liminf \psi(x) < 1$. Since we have already shown that $\psi(p^k) = 1$, we merely need to prove that $\psi(x) \leq 1$ and the strict inequality holds for some $x$.

**Lemma 5** *Let $f(x) = \{\log(x+1)/2\}/\log x$. Then $f(x)$ is increasing for $x > 1$.*

**Proof:** $f'(x) = \{\frac{\log x}{x+1} - \frac{\log(x+1)/2}{x}\}/(\log x)^2$ has the same sign as its numerator, which has the same sign as $g(x) = x\log x + (x+1)\log 2 - (x+1)\log(x+1)$. $g'(x) = \log x + \log 2 - \log(x+1) = \log\{2x/(x+1)\} > 0$ for $x > 1$, so $g(x)$ is increasing for $x > 1$. Since $g$ is continuous for $x \geq 1$ and $g(1) = 0$, it follows that $g(x) > 0$, and hence $f'(x) > 0$ as well, for $x > 1$. Thus $f$ is increasing for $x > 1$.

∎

We use this lemma to show

**Lemma 6** *$\psi(x) < 1$ for $1 < x < p$.*

**Proof:** By (8), if $1 \leq x \leq p$, $\phi(x) = x(x+1)/2$. We can thus write

$$\psi(x) = \phi(x)/x^\theta = x^{\frac{x(x+1)/2}{\log_p(p(p+1)/2)}} = \left(\frac{e^{f(x)}}{e^{f(p)}}\right)^{\log x},$$

where f is defined as in the preceding lemma. The monotonicity of $f$ implies that $\psi(x) < 1$ if $1 < x < p$.

∎

We are now in a position to prove

**Theorem 7** *$\psi(x) \leq 1$ for all $x$.*

**Proof:** We use induction. Assume $\psi(x) \leq 1$ for $x \leq p^k$ and suppose $p^k \leq x \leq p^{k+1}$. Then $x$ may be written in the form $ap + b$, with $a < p^k, b \leq p$. Since, by lemma 3, $\phi(ap+b) = P\phi(a) + T(a)\phi(b)$ and, by the induction, $\phi(b) \leq b^\theta$,

$$\psi(ap+b) \leq \frac{P\phi(a) + T(a)b^\theta}{(pa+b)^\theta} = \frac{\phi(a) + T(a)(b/p)^\theta}{(a+b/p)^\theta}.$$

This is of the form $g(t) = \{\phi(a) + T(a)t^\theta\}/(a+t)^\theta$ with $0 \leq t \leq 1$. We need only show $g(t) \leq 1$. First observe that $g'(t) = \theta\{aT(a)t^{\theta-1} - \phi(a)\}/(a+t)^{\theta+1}$, so there exists some $t_0$ such that $g'(t) < 0$ for $t < t_0$ and $g'(t) > 0$ for $t > t_0$. Therefore, any maximum for $g$ on $[0,1]$ must occur at an endpoint. But $g(0) = \phi(a)/a^\theta = \psi(a)$, $g(1) = \{\phi(a) + T(a)\}/(a+1)^\theta = \phi(a+1)/(a+1)^\theta = \psi(a+1)$ and both $\psi(a) \leq 1$ and $\psi(a+1) \leq 1$ by the induction hypothesis.

∎

We can do more than show that $\liminf \psi(x) < 1$. For any x, we can easily find some $x^* > x$ such that $\psi(x^*) < \psi(x)$. Towards this end, we prove the following two lemmas.

**Lemma 8** *If $y \leq (2p-1)/3$, then $\sum_{j=0}^{y-1} T(px+j) \leq \sum_{j=0}^{y-1} T(px-j-1)$.*

**Proof:** From (7), we see that $T(px+j) = T(x)(j+1)$ and thus

$$\sum_{j=0}^{y-1} T(px+j) = T(x)\sum_{j=0}^{y-1}(j+1) = T(x)y(y+1)/2.$$

Comparing the digits of $x$ and $px-j-1$, we see that all but one non-zero digit of $x$ occurs in $px-j-1$. Corresponding to the digit $x^*$ of $x$ that does not appear in $px-j-1$ is $x^*-1$, while $px-j-1$ has a digit $p-j-1$ not occurring in $x$. Using (7), it is clear that $T(px-j-1) \geq T(x)(p-j)/2$.

Therefore, $\sum_{j=0}^{y-1} T(px-j-1) \geq T(x)\sum_{j=0}^{y-1}(p-j)/2 = T(x)\{yp - y(y-1)/2\}/2$. The inequality $\sum_{j=0}^{y-1} T(px+j) \leq \sum_{j=0}^{y-1} T(px-j-1)$ will hold if $T(x)y(y+1)/2 \leq T(x)\{yp-y(y-1)/2\}/2$, which holds if $y \leq (2p-1)/3$. ∎

**Lemma 9** *If* $y \leq (2p-1)/3$, *then* $\phi(px+y)+\phi(px-y) \leq 2\phi(px)$.

**Proof:** $\phi(px+y) = \phi(px)+\sum_{j=0}^{y-1} T(px+j)$ and $\phi(px-y) = \phi(px)-\sum_{j=0}^{y-1} T(px-j-1)$. Adding these together and using the previous lemma gives the result desired. ∎

We are now prepared to show how, given any $x$, there is some $x^*$ such that $\psi(x^*) < \psi(x)$.

**Theorem 10** *If* $0 < y \leq (2p-1)/3$, *then* $\min\{\psi(px+y), \psi(px-y)\} < \psi(x)$.

**Proof:** Suppose the assertion is false. Then there is some pair $x, y$ with $y \leq (2p-1)/3$ such that $\phi(px+y)/(px+y)^\theta \geq \phi(x)/x^\theta$ and $\phi(px-y)/(px-y)^\theta \geq \phi(x)/x^\theta$. But then $\phi(px+y)+\phi(px-y) \geq \phi(x)/x^\theta\{(px+y)^\theta + (px-y)^\theta\}$ and therefore, using the preceding lemma,

$$\phi(x)/x^\theta\{(px+y)^\theta+(px-y)^\theta\} \leq 2\phi(px) = 2P\phi(x),$$

which implies that $(1+\frac{y}{px})^\theta + (1-\frac{y}{px})^\theta \leq 2$, which is a contradiction since $0 < \frac{y}{px} < 1$. ∎

This theorem is a generalization of what Harborth proved for $p=2$, which was used to construct a sequence on which $\psi$ was strictly decreasing and for which he conjectured $\psi$ approached its lim inf. Analogously, we can define sequences $\{x_r\}, \{y_r\}$ by letting $x_0 = 1$ and $x_{r+1} = px_r \pm y_r$ where $0 < y_r < p$ is chosen to minimize $\psi(x_{r+1})$. *We conjecture that* $\psi(x_r) \to \liminf \psi(x)$.

# 3 Extensions of $\psi$

$\psi$ can be extended canonically to a continuous function on $\Re^+$ and that extension suggests an alternate interpretation of the last conjecture. The extension of $\psi$ to $\Re^+$ comes in two stages. First we extend $\psi$ to $Q^+$.

**Definition:** Let $x$ be a positive rational. Choose some integer $k$ such that $p^k x$ is an integer. We define $\psi(x)$ to be equal to $\psi(p^k x)$, which has previously been defined.

This definition is valid because we already know that $\psi(p^k x)$ is independent of $k$ as long as $p^k x$ is an integer.

We can look at this definition as follows. Write down a finite sequence of $p$-ary digits and place a radix point somewhere, yielding a number $x$. If all the digits after the radix point are 0s, then $x$ is an integer and we already know that $\psi(x)$ is independent of exactly where the radix point is placed, so that the value of $\psi$ really depends only on the sequence. By extending the domain of $\psi$, we have merely allowed the placement of the radix point anywhere.

We now extend the definition of $\psi$ to $\Re^+$ in the natural way.

**Definition:** If $x \in \Re^+$, choose a sequence $x_r \to x$ of rationals and let $\psi(x) = \lim \psi(x_r)$.

We can look at this extension in the spirit of allowing the sequences described above to be infinite as well as finite. To prove the validitiy of this extension, we have to show that if two sequences start off the same way and we apply $\psi$ to both, the resulting values are close to one another. This is the end towards which the next few lemmas are directed.

**Lemma 11** *If $x$ is a positive integer, then $\psi(x) - \theta/x \le \psi(x+1) \le \psi(x) + 1/x^{\theta-1}$.*

**Proof:** $\phi(x+1) = \phi(x) + T(x) \le \phi(x) + (x+1)$. Therefore, $\psi(x+1) \le (\phi(x) + x + 1)/(x+1)^\theta \le \phi(x)/x^\theta + 1/(x+1)^{\theta-1} \le \psi(x) + 1/x^{\theta-1}$.

On the other hand, $\psi(x+1) = \psi(x) - \{\phi(x)/x^\theta - \phi(x+1)/(x+1)^\theta\} \ge \psi(x) - \{\phi(x)/x^\theta - \phi(x)/(x+1)^\theta\} = \psi(x) - \psi(x)\{1 - 1/(1+1/x)^\theta\} \ge \psi(x) - \theta/x$. ∎

**Corollary 12** *Let $x \ge p^s$ be a positive integer. Then $\psi(x) - \theta/p^s \le \psi(x+1) \le \psi(x) + \{2/(p+1)\}^s$.*

**Lemma 13** *Consider sequences $\{x_r\}, \{y_r\}$ of integers where $x_0 \ge p^s, 0 \le y_r < p$ and $x_{r+1} = px_r + y_r$. Then*

$$\psi(x_0) - \theta/p^s \le \psi(x_n) \le \psi(x_0) + \left(\frac{p+1}{p-1}\right)\left(\frac{2}{p+1}\right)^s.$$

**Proof:** First observe that

$$\psi(x_{r+1}) - \psi(x_r) = \phi(px_r + y_r)/(px_r + y_r)^\theta - \phi(x_r)/x_r^\theta. \tag{10}$$

From (10)

$$\begin{aligned}\psi(x_{r+1}) - \psi(x_r) &\ge \phi(px_r)/(px_r + y_r)^\theta - \phi(x_r)/x_r^\theta \\ &= \{\phi(x_r)/x_r^\theta\}\{1/(1+\frac{y_r}{px_r})^\theta - 1\}.\end{aligned}$$

Since $y_r < p$ and $x_r \ge p^{s+r}$, it follows that

$$\psi(x_{r+1}) - \psi(x_r) \ge 1/\left(1 + \frac{p-1}{p^{s+r+1}}\right)^\theta - 1 \ge -\theta(p-1)/p^{s+r+1}.$$

Thus

$$\psi(x_n) \ge \psi(x_0) - \sum_{r=0}^{\infty} \theta(p-1)/p^{s+r+1} = \psi(x_0) - \theta/p^s.$$

For the other half of the inequality, since $\phi(px_r+y_r) = \phi(px_r)+T(x_r)\phi(y_r)$, using (10), we get

$$\psi(x_{r+1}) - \psi(x_r) = \frac{\phi(px_r) + T(x_r)\phi(y_r)}{(px_r + y_r)^\theta} - \phi(x_r)/x_r^\theta \leq \frac{T(x_r)\phi(y_r)}{(px_r + y_r)^\theta} \leq$$

$$\frac{T(x_r)}{x_r{}^\theta} \leq (x_r+1)/(x_r+1)^\theta \leq x_r/x_r{}^\theta \leq p^{s+r}/(p^{s+r})^\theta = \left(\frac{2}{p+1}\right)^{s+r}.$$

Thus

$$\psi(x_n) \leq \psi(x_0) + \sum_{r=0}^{\infty}\left(\frac{2}{p+1}\right)^{s+r} = \psi(x_0) + \left(\frac{p+1}{p-1}\right)\left(\frac{2}{p+1}\right)^s.$$

∎

We are now in a position to show that $\psi$, restricted to the positive rationals, is a uniformly continuous function on any interval bounded away from 0.

**Lemma 14** *Let $a, \epsilon > 0$. Then there exists some $\delta > 0$ such that if $x, x' \geq a$ are rationals with $|x' - x| < \delta$, then $|\psi(x') - \psi(x)| < \epsilon$.*

**Proof:** First choose $s$ so that

$$\max\{\left(\frac{2}{p+1}\right)^s, \theta/p^s, \left(\frac{p+1}{p-1}\right)\left(\frac{2}{p+1}\right)^s\} < \epsilon/3.$$

Next, choose $t \geq s$ such that $p^t a \geq p^s$ and let $\delta = 1/p^t$. Suppose $|x' - x| < \delta$. Without loss of generality, we may assume $x < x'$. Then $0 < p^t x' - p^t x < 1$ and we may consider two cases:

*Case 1: There exists an integer $x_0$ such that $p^s \leq x_0 \leq p^t x < p^t x' < x_0 + 1$.* Then there are sequences $\{x_r\}, \{y_r\}$ as in lemma 13 such that $x_n = p^{t^*} x$ for some $t^*$ and thus $|\psi(x) - \psi(x_0)| = |\psi(x_0) - \psi(x_n)| < \epsilon/3$. Similarly, it can be shown that $|\psi(x') - \psi(x_0)| < \epsilon/3$, so $|\psi(x') - \psi(x)| < 2\epsilon/3 < \epsilon$.

*Case 2: There is an integer $x_0$ such that $p^s \leq x_0 < p^t x < x_0 + 1 \leq p^t x'$.* The same argument as used in the other case shows that $|\psi(x) - \psi(x_0)| < \epsilon/3$ and $|\psi(x') - \psi(x_0+1)| < \epsilon/3$, while lemma 11 shows that $|\psi(x_0+1) - \psi(x_0)| < \epsilon/3$ and we can again conclude that $|\psi(x') - \psi(x)| < \epsilon$.

∎

The uniform continuity of $\psi$ on all intervals $[a, \infty) \subset Q$ implies that the canonical extension of $\psi$ to $\Re^+$ is also uniformly continuous on all intervals $[a, \infty) \subset Q$. The lemmas used in the proof also enable us to estimate any value of $\psi$ to any desired accuracy. In particular, we can estimate $\inf \psi$ with the following immediate corollary of lemma 13.

**Corollary 15** *If $m_s = \min_{p^s \leq n \leq p^{s+1}, n \in Z} \psi(n)$, then $\inf \psi(x) \geq m_s - \theta/p^s$.*

# 4 Sets of Primes

Generalizations of Lucas' Theorem have been used to show that, given any arbitrary integer, almost all binomial coefficients are divisible by that integer. As a special case (where the integer is squarefree), we know that, given a finite set of primes, almost all binomial coefficients are divisible by all the primes in that set. Using the bound $\psi(x) \leq 1$, we can determine how far we must look before we are guaranteed to find at least one such binomial coefficient.

**Theorem 16** *Let $\mathcal{P}$ be a set of primes, $|\mathcal{P}| = N$, let $p^*$ be the largest prime in $\mathcal{P}$ and let $\theta^* = \{\log p^*(p^*+1)/2\}/\log p^*$. If $x \geq (2N)^{1/(2-\theta^*)}$, then there is at least one binomial coefficient $\binom{n}{m}$, with $n < x$, divisible by all elements of $\mathcal{P}$.*

**Proof:** Since $\psi_p(x) \leq 1$, $\sum_{p\in\mathcal{P}} \phi_p(x) \leq \sum_{p\in\mathcal{P}} x^{\theta_p} \leq N x^{\theta^*}$ and this is a bound on the number of binomial coefficients that fail to be divisible by at least one prime in $\mathcal{P}$. The total number of binomial coefficients under consideration is $x(x+1)/2 \geq x^2/2$. Thus, there will be at least one binomial coefficient divisible by every prime in $\mathcal{P}$ if $x \geq (2N)^{1/(2-\theta^*)}$.

∎

# 5 The Generating Function

Our last result is a product representation for the generating function for T:

$$\sum_{n=0}^{\infty} T(n)x^n = \prod_{i=0}^{\infty} \frac{1-(p+1)x^{p^{i+1}} + px^{p^i(p+1)}}{(1-x^{p^i})^2}. \tag{11}$$

Since $T(n) = \prod(n_i+1)$ if $n = \sum n_i p^i$, $n_i < p$, we may write $\sum T(n)x^n = \prod_{i=0}^{\infty}\{1+2x^{p^i}+3x^{2p^i}+\ldots+px^{(p-1)p^i}\}$.

If we write an individual factor as $g(p^i)$, where $g(t) = 1+2t+3t^2+\ldots+pt^{p-1}$, we see that $g(t) = G'(t)$, where $G(t) = 1+t+t^2+\ldots+t^p = (1-t^{p+1})/(1-t)$. Differentiating the quotient yields $g(t) = (1-(p+1)t^p + pt^{p+1})/(1-t)^2$, which gives (11) when we replace $t$ by $x^{p^i}$. Both the series and the product converge for $|x| < 1$.

# 6 Conclusion

We have taken results concerning binomial coefficient parity and generalized them to binomial coefficients modulo an arbitrary prime. To summarize, we have shown

- $\phi(x) \leq x^\theta$.
- $\phi(x) > x^\theta/P$, and we have a conjecture about a sequence converging to $\liminf \phi(x)/x^\theta$.
- We have extended $\psi$ as a continuous function on $\Re^+$.
- We have seen how to estimate $\psi(x)$ to whatever precision desired.
- We have obtained a product representation for the generating function for $T(x)$.

# References


[1] J. Coquet. *A summation formula related to the binary digits.* Invent. Math. **73** (1983), 107–115.

[2] N.J. Fine. *Binomial coefficients modulo a prime.* Amer. Math. Monthly **54** (1947), 589–592.

[3] J.W.L. Glaisher. *On the residue of a binomial-theorem coefficient with respect to a prime modulus.* Quart. J. Pure. App. Math. **30** (1899), 150-156.

[4] H. Harborth. *Number of odd binomial coefficients.* Proc. Amer. Math. Soc. **63** (1977), 19-22.

[5] M.D. McIlroy. *The number of 1's in binary integers: bounds and extremal properties.* SIAM J. Comput. **3** (1974) 255-261.

[6] David Singmaster. *Notes on binomial coefficients I– A generalization of Lucas' congruence.* J. London Math. Soc. **8** (1974), 545-548.

[7] David Singmaster. *Notes on binomial coefficients II – The least n such that $p^e$ divides an r-nomial coefficient of rank n.* J. London Math. Soc. **8** (1974), 549-554.

[8] David Singmaster, *Notes on binomial coefficients III – Any integer divides almost all binomial coefficients.* J. London Math. Soc. **8** (1974), 555-560.

[9] A.H. Stein. *Exponential sums related to binomial coefficient parity.* Proc. Amer. Math. Soc. **80** (1980), 526-530.

[10] A.H. Stein. *Exponential sums of an iterate of the binary sum-of-digit function.* Indiana Univ. Math. J. **31** (1982), 309-315.

[11] A.H. Stein. *Exponential sums of sum-of-digit functions.* Illinois J. Math. **30** (1986), 660-675.

[12] A.H. Stein. *Exponential sums of digit counting functions.* (to appear).

[13] K. Stolarsky, *Power and exponential sums related to binomial coefficient parity.* SIAM J. Appl. Math. **32** (1977), 717-730.


The University of Connecticut
32 Hillside Avenue
Waterbury, CT 06710

## POSITIVE CHARACTERISTIC CALCULUS AND ICEBERGS

Moss E. Sweedler[1]
Department of Mathematics
Cornell University
Ithaca New York 14853

This overview of joint work with Mitsuhiro Takeuchi begins with a positive characteristic analog to the elementary calculus result: **closed if and only if exact.**[2] The positive characteristic analog, **p-closed if and only if exact**, has applications in characteristic zero as well as positive characteristic. We outline how it yields a new proof that dX/X does not have a rational function integral in any characteristic. Then we describe how **p-closed if and only if exact** helps with symbolic integration of closed one-forms in positive characteristic. The section, **POSITIVE CHARACTERISTIC ICEBERGS**, is about the "big picture" and how we arrived at the result **p-closed if and only if exact.** The last section, **THE GOOD THE BAD AND THE UGLY**, is a critique. [1] is an overview of other aspects of our joint work. The majority of the work itself appears in [2].[3] Since this is an expository article, results will be stated to facilitate presentation rather than be put in their most general form.

### p-CLOSED IF AND ONLY IF EXACT

Throughout, A is either $R[X_1,\cdots,X_n]$ or $R(X_1,\cdots,X_n)$ and $\omega = \Sigma\, a_i\, dX_i$ is a **one-form** over A, meaning that the $\mathbf{a_i}$'s lie in A. As usual $\omega$ is called:

**EXACT**: if there is $a \in A$ with $\omega = \Sigma\, ((\partial/\partial X_i)a)\, dX_i$ ( $\equiv$ **grad** a )

**CLOSED**: if $(\partial/\partial X_i)a_j = (\partial/\partial X_j)a_i$ for $i \neq j$

Whether A is $R[X_1,\cdots,X_n]$ or $R(X_1,\cdots,X_n)$, **exact implies closed** because the partials $\partial/\partial X_i$ and $\partial/\partial X_j$ commute. In calculus we learn that if R is **R** or **C** and A is $R[X_1,\cdots,X_n]$ then $\omega$ is exact if and only if $\omega$ is closed. The usual proof of **closed implies exact** is to integrate, one variable at a time. This proof applies as long as R is a field of characteristic zero and A is

---

[1]Supported in part by the Japan Society for the Promotion of Science, the Alexander Von Humboldt Foundation, the National Science Foundation and IBM.

[2]This and the section: **POSITIVE CHARACTERISTIC ICEBERGS** will be understandable with little more than a calculus background.

[3]However, some of the results mentioned here were discovered after [2].

$R[X_1,\cdots,X_n]$. If A is $R(X_1,\cdots,X_n)$ or R is a field of positive characteristic, then closed does not imply exact. What does imply exact? We shall give an answer which applies equally well to $R[X_1,\cdots,X_n]$ or $R(X_1,\cdots,X_n)$ when R is a field of positive characteristic. Here is the kind of question we shall be able to easily answer:

1 QUESTION: Let $f(X_1,X_2) = X_1^3X_2^2 + X_1^4X_2 + X_1^5$, let $\omega = f\,dX_1 - f\,dX_2$ and say the characteristic is 3. $\omega$ is closed. Is $\omega$ exact?

2 DEFINITION: For $0 < p \in \mathbb{Z}$, $\omega$ is called **p-closed** if:

a. $\omega$ is closed
b. $(\partial/\partial X_i)^{p-1} a_i = 0$ for all i.

3 p-CLOSED IF AND ONLY IF EXACT THEOREM: Suppose R is a field of positive characteristic p and A is $R[X_1,\cdots,X_n]$ or $R(X_1,\cdots,X_n)$. One-forms over A are p-closed if and only if they are exact.

SKETCH OF PROOF: As mentioned above, exact implies closed because the partials commute. Exact implies the p - 1 power vanishing condition, (2,b), because in positive characteristic p: $(\partial/\partial X_i)^p = 0$. This is easily verified for the polynomial ring which implies that $(\partial/\partial X_i)^p = 0$ for the rational function field. Thus, **exact implies p-closed.** The proof that **p-closed implies exact** for $A = R[X_1,\cdots,X_n]$ is essentially the standard proof from calculus that closed implies exact, taking a little care to insure that after integrating with respect to each variable the p-closed condition is preserved.

**p-closed implies exact** for $A = R(X_1,\cdots,X_n)$ comes from the case $A = R[X_1,\cdots,X_n]$ and a trick to clear denominators. Since there are only a finite number of $\mathbf{a_i}$'s in: $\omega = \Sigma\, a_i\, dX_i$ and the $\mathbf{a_i}$'s lie in $R(X_1,\cdots,X_n)$, there is a polynomial b in $R[X_1,\cdots,X_n]$ which clears denominators in the sense that all the products, $ba_i$, are polynomials in $R[X_1,\cdots,X_n]$. Thus the products, $b^pa_i$, are in $R[X_1,\cdots,X_n]$ and the one-form $\gamma$ which is defined as $b^p\omega$ is a one-form over $R[X_1,\cdots,X_n]$. Derivations vanish on $p^{th}$ powers in characteristic p. Thus $p^{th}$ powers act like constants as far as derivations are concerned. We sometimes refer to $p^{th}$ powers as **derivation-constants**. Because $b^p$ is a derivation-constant, $\gamma$ has the same properties as $\omega$, scaled by $b^p$, with respect to derivations. In particular, $\gamma$ is a p-closed one-form over $R[X_1,\cdots,X_n]$. Since we are considering **p-closed implies exact** to be proved for $A = R[X_1,\cdots,X_n]$ , there is a polynomial a in $R[X_1,\cdots,X_n]$ with $\gamma = \mathbf{grad}\ a$ . Since "**grad**" is based on derivations, $\omega = \mathbf{grad}\ (a/b^p)$ . ■

Let us apply the theorem to the question posed in (1). $(\partial/\partial X_1)^2 f = 2X_1^3 \neq 0$ . Hence, $\omega$ is not p-closed, or 3-closed in this case, and so is not exact.

4 EXAMPLE: In positive characteristic p let $\omega_1 = X^{p-1}\,dX$ and $\omega_2 = dX/X$ . Since $\omega_1 = X^p\omega_2$ they differ by a multiplicative derivation-constant. Thus $\omega_1$ and $\omega_2$ will have the same properties with respect to: exactness, closedness and p-closedness. Both are closed because they involve only one variable. Neither are p-closed:

$\omega_1$: $\quad (\partial/\partial X)^{p-1}\,(X^{p-1}) = (p-1)! \equiv -1 \bmod p$

$\omega_2$: $\quad (\partial/\partial X)^{p-1}\,(1/X) = (p-1)!\,/\,X^p$

Hence they are not exact, giving an extremely easy proof that dX/X has no rational function integral in positive characteristic. This leads to a new algebraic proof that dX/X has no rational function integral in characteristic zero. Consider R(X) where R is a field of characteristic zero. Suppose there were $u(X),v(X) \in R[X]$ with:

$$\mathbf{grad}\ u/v = dX/X \qquad (5)$$

Assume for the moment that u(X) and v(X) have integer coefficients. Choose a prime number p not dividing all the coefficients of v. Reduce (5) mod p to get:

$$\mathbf{grad}\ \overline{u}\,/\,\overline{v} = dX/X$$

which contradicts the fact that in positive characteristic dX/X has no rational function integral. Of course, u(X) and v(X) will not have integral coefficients in general. However the coefficients can always be assumed to lie in a subring of R of the form $\mathbb{Z}[Y_1,\cdots,Y_s,\alpha]$ where $Y_1,\cdots,Y_s$ are algebraically independent over $\mathbb{Q}$ and $\alpha$ is integral over $\mathbb{Z}[Y_1,\cdots,Y_s]$ . Instead of reducing (5) mod a prime number, reduce mod a prime ideal in $\mathbb{Z}[Y_1,\cdots,Y_s,\alpha]$ which does not contain all the coefficients of v(X), but does contain an integral prime.

## POSITIVE CHARACTERISTIC SYMBOLIC INTEGRATION OF CLOSED ONE-FORMS

p-closed consists of the closed condition, (2,a), and the p - 1 power vanishing condition, (2,b). In positive characteristic p, when one-forms are closed but do not satisfy the p - 1 power vanishing condition, we can apply our results to obtain the rational and irrational parts of the integral. The non-zeroness of the $\mathbf{b_i}$'s, defined below, is the measure of $\omega$ not satisfying condition (2,b). Therefore it is not surprising that the $\mathbf{b_i}$'s split $\omega$ into parts with rational and irrational integrals.

6 INTEGRATION THEOREM: A is either $R[X_1,\cdots,X_n]$ or $R(X_1,\cdots,X_n)$ where R is a field of positive characteristic p. Let $\omega = \Sigma\, a_i\, dX_i$ be a closed, but not necessarily p-closed, one-form over A. Let $b_i = (\partial/\partial X_i)^{p-1} a_i$ and set:

$$\omega_{rat} = (\Sigma\, a_i + b_i X_i^{p-1})\, dX_i \qquad \omega_{irrat} = -\Sigma\, b_i X_i^{p-1}\, dX_i$$

so that $\omega = \omega_{rat} + \omega_{irrat}$.

a. $\omega_{rat}$ is p-closed, hence exact. $\omega_{rat}$ is the rational part of the integral of $\omega$.

b. $\omega_{irrat} = \mathbf{grad}\,(\Sigma\, b_i X_i^p \log X_i)$. $\omega_{irrat}$ is the irrational part of the integral of $\omega$.

c. The $\mathbf{b_i}$'s lie in $R[X_1{}^p,\cdots,X_n{}^p]$ or $R(X_1{}^p,\cdots,X_n{}^p)$ according to the choice of A. Thus if R is a perfect field, the $\mathbf{b_i}$'s are $p^{th}$ powers.

SKETCH OF PROOF: All partials vanish on the $\mathbf{b_i}$'s because:

For $i = j$ : $(\partial/\partial X_i)b_i = 0$ < since $(\partial/\partial X_i)^p = 0$ >

For $i \neq j$:

$(\partial/\partial X_j)b_i$ = < definition of $b_i$ > $(\partial/\partial X_j)(\partial/\partial X_i)^{p-1} a_i$

= < partials commute > $(\partial/\partial X_i)^{p-1}(\partial/\partial X_j)a_i$

= < $\omega$ is assumed closed > $(\partial/\partial X_i)^{p-1}(\partial/\partial X_i)a_j$

= 0 < since $(\partial/\partial X_i)^p = 0$ >

Using the fact that all partials vanish on the $\mathbf{b_i}$'s, it is simple to check that $\omega_{irrat}$ is closed. Since $\omega_{rat} = \omega - \omega_{irrat}$, $\omega_{rat}$ is closed. It is easy to check that $(\partial/\partial X_i)^{p-1}(b_i X_i^{p-1}) = -b_i$. Thus $\omega_{rat}$ is p-closed as claimed in part a. Using the fact that the partials vanish on the $\mathbf{b_i}$'s it is easy to check that $\omega_{irrat}$ has the integral asserted in part b. The fact that all the partials vanish on the $\mathbf{b_i}$'s implies that they lie where asserted in part c. ■

## POSITIVE CHARACTERISTIC ICEBERGS

The usual calculus result, **closed if and only if exact**, is the protruding tip of the Poincaré lemma iceberg which lies in the DeRham cohomology ice field. Algebraic DeRham cohomology exists in positive characteristic, but is ill behaved. For example, the algebraic Poincaré lemma, using algebraic DeRham cohomology, sinks in positive characteristic. The ill mannered nature of algebraic DeRham cohomology in positive characteristic is among the motivating factors behind the development of some of the fancy cohomology theories of algebraic geometry.

**p-closed if and only if exact** is the protruding tip of an iceberg in a newly discovered ice field. Mitsuhiro Takeuchi and the author have been exploring this new field since 1982. We arrived at **p-closed if and only if exact** from underwater with the aid of a computer, rather than noticing it from above. In the spring of 1986 we recognized the possibility that p-closed might imply exact.[4] We tried to prove that p-closed implies a homological condition which we knew implied exact. At first the problem looked out of reach and we started working examples. By hand we could only work small examples in characteristic 2 or 3. Using IBM's algebraic computation language, Scratchpad II, we were able to work examples up to characteristic 17. For the homological condition to be satisfied, a certain computation had to yield zero. It always did. The computation involved derivations of a commutative ring in positive characteristic. To get more information we tried running the program without the commutativity assumption on the variables. The answer was no longer zero but one could see how the terms would cancel with commutativity. Looking at the non-commutative case provided the key to our first proof of p-closed implies exact. The proof went:

---

[4] The other direction is easy.

p-closed => the terms always cancel

=> the homological condition => exact

At an early presentation of **p-closed if and only if exact**, Don Passman made the lovely[5] and embarrassing[6] observation that the standard proof from calculus could be supplemented with a little bookkeeping to give p-closed implies exact.

Floating in the new ice field are many algebra analogs to results from differential geometry. Here are a few of the major analogies:

| | |
|---|---|
| closed if and only if exact | p-closed if and only exact |
| Poincaré lemma | a Poincaré type lemma, [2, (8.29)] |
| DeRham cohomology | a new cohomology theory [2, section 8] |
| Frobenius theorem on integral submanifolds | Jacobson's intermediate field theory for purely inseparable exponent one field extensions [2, section 6] |

## THE GOOD THE BAD AND THE UGLY

In differential geometry one considers one-forms as a module over the ring of $C^\infty$ functions and constructs the Grassman or exterior algebra. This gives the underlying complex of the De-Rham complex. The degree zero, one and two stages of the DeRham complex, and the intermediate maps, are used in the differential ideal formulation of the Frobenius theorem on integral submanifolds. This formulation of the Frobenius theorem is analogous to a new covariant, functorial formulation of Jacobson theory which is part of the new ice field, [2, section 6]. A *twenty-five words or less* description of the analogy: Jacobson's intermediate fields correspond to subalgebras of $C^\infty$ functions which are constant along integral submanifolds.

---

5Lovely, because it shows **p-closed if and only if exact** is an elementary result which could have been observed and proved years ago.

6Embarrassing, because we had gone through such contortions to get the result.

The exterior algebra is a graded skew commutative differential algebra. The new cohomology is based on a complex arising from a graded commutative[7] differential algebra, $(T,t)$. Suppose we are dealing with a commutative algebra B over a base ring R. The algebraic DeRham complex is B in degree zero and $T_0$ is B. The algebraic DeRham complex is the Kaehler module of B over R in degree one and $T_1$ is the Kaehler module of B over R. In each case the map from degree zero to degree one is the usual differential from B to its Kaehler module, [2, (1.12,a), (2.6)]. Those who are familiar with the construction of the Kaehler module as $I/I^2$, where I is the kernel of the multiplication map from $B \otimes_R B$ to B, may be intrigued to learn that $T_2$ is $I/I^3$. (For the exact hypotheses of the $I/I^3$ result, see [2, (2.14)].) However $T_3$ is not $I/I^4$, [2, example following (2.17)].

The derivation, t, does not have square zero. It has $p^{th}$ power zero. For each i between 1 and p - 1 there is a complex:

$$7,i \quad T_0 \xrightarrow{t^i} T_i \xrightarrow{t^{p-i}} T_p \xrightarrow{t^i} T_{p+i} \xrightarrow{t^{p-i}} T_{2p} \xrightarrow{t^p} \cdots$$

When B is the polynomial ring over a field of positive characteristic, these complexes are acyclic, [2, (8.29)]. This gives p - 1 Poincaré lemmas. The case $i = 1$ is the case which looks most like the traditional characteristic zero Poincaré lemma. **p-closed if and only if exact** is the consequence of the vanishing homology in degree one of (7,i), for i=1, in the same way that in characteristic zero **closed if and only if exact** is the consequence of the vanishing homology in degree one for the algebraic DeRham complex of the polynomial ring.

After the good news comes the bad: **THE COMPLEXES (7,I) ARE ACYCLIC TOO OFTEN.** The homology does not reflect the underlying geometry. The homology vanishes for examples where, geometrically speaking, it should not. For example if R is any field of positive characteristic and B is $R[X]_X$, which represents the punctured plane, the degree one homology should be one dimensional. Unfortunately it is zero. This gives a **p-closed if and only if exact** result for one-forms over $R[X]_X$. I would rather have the homology reflect the geometry. [2, (8.29)] gives a p-basis criterion for when the homology vanishes.

---

[7] Honest commutative not skew commutative.

The ad hoc construction of the complex(es), (7,i), is ugly. The acyclicity of the complexes is an easy consequence of [2, (8.20)] which proves that a certain module is free and gives information about (T,t). [2, (8.20)] is a fundamental result and is not ugly.[8]

## REFERENCES

[1] M. Sweedler. Introduction to the algebraic theory of positive characteristic differential geometry, Lecture Notes in Mathematics, 1146, Springer-Verlag, (1985), 317-324.

[2] M. Sweedler and M. Takeuchi. From differential geometry to differential algebra: analogs to the Frobenius theorem and Poincaré lemma, IBM Research Report RC 12082, 198 pages. To appear, Springer Lecture Notes in Mathematics.

---

[8] (8.20) gave us a hard time. One morning we despaired of proving it. That afternoon we got it.

## Arithmetic of Certain Algebraic Surfaces over Finite Fields*

Noriyuki Suwa and Noriko Yui

### Contents

Introduction

1. Crystalline cohomology and de Rham-Witt complex
2. The Tate conjecture and the Artin-Tate formula
3. Abelian surfaces
4. Kummer surfaces
5. Fermat surfaces

### Introduction.

Let $K$ be a number field. Let $\zeta_K(s) = \prod_{\mathfrak{p}} (1-N\mathfrak{p}^{-s})^{-1} = \sum_{\mathfrak{a}} N\mathfrak{a}^{-s}$ denote the Dedekind zeta-function of $K$. It is known in some cases and is conjectured in general that the order of poles (or zeros) and the special values of $\zeta_K(s)$ at $s = n$ (integer $\geq 0$) yield some valuable information about the field $K$, such as the class number, the regulator, the discriminant, and the higher $K$-groups and the higher regulators (cf. Lichtenbaum [18]).

Now let $k = \mathbb{F}_q$ denote a finite field of characteristic $p > 0$ and let $\bar{k}$ be its algebraic closure. Let $X$ be a smooth projective variety of dimension $d$ over $k$. Let $\zeta(X;s) = \prod_{x \in X^0} (1 - Nx^{-s})$ be the Hasse-Weil zeta-function of $X$. Then

$$\zeta(X;s) = Z(X;q^{-s}) ,$$

where

$$Z(X;T) = \exp( \sum_{n=1}^{\infty} \frac{N_n}{n} T^n )$$

---

*) This is an extended version of a talk presented by N. Yui at New York Number Theory Seminar at Graduate Center, CUNY, on October 20, 1987.

This work was partially supported by the Natural Sciences and Engineering Research Council of Canada (NSERC) under Grant #A 8566 and #A 9451.

is the congruence zeta-function of $X$. (Here $N_n = |X(\mathbb{F}_{q^n})|$ stands for the cardinality of the $\mathbb{F}_{q^n}$-rational points on $X$.) It is known that $Z(X;T)$ is a rational function of the form

$$Z(X;T) = \prod_{i=0}^{2d} P_i(X;T)^{(-1)^{i+1}}$$

where

$$P_i(X;T) = \begin{cases} \det(1-\Phi T; H^i(X_{\bar{k}}, \mathbb{Q}_\ell) & \text{for } \ell \text{ prime} \neq p = \mathrm{char}(k) \\ \det(1-\Phi T; H^i(X/W)_K) \end{cases}$$

is the characteristic polynomial of the geometric Frobenius, $\Phi$, on the $\ell$-adic étale cohomology group $H^i(X_{\bar{k}}, \mathbb{Q}_\ell)$ for any prime $\ell \neq p$, and on the crystalline cohomology group $H^i(X/W)_K$.

One knows that $P_i(X;T) \in \mathbb{Z}[T]$ with $\deg P_i = B_i(X)$ (the i-th Betti number of $X$), and furthermore, that the reciprocal roots $\alpha_{ij}$ of $P_i(X;T) = 0$ have the absolute value $q^{i/2}$ in any embedding into $\mathbb{C}$ (Deligne [5]).

Therefore, the order, $\rho_i$, of poles at $s = i$ of $\zeta(X:s)$ is equal to the multiplicity of $q^i$ as a reciprocal root of $P_{2i}(X;T) = 0$. In these circumstances, analysis of the special values of $\zeta(X;s)$ at $s = i$ may be reduced to the evaluation of $P_{2i}(X;T)/(1-q^iT)^{\rho_i}$ at $T = q^{-i}$ and those of $P_j(X;1/q^i)$ for $j \neq 2i$.

Let $X$ be a smooth projective curve over $k = \mathbb{F}_q$. Then the congruence zeta-function of $X$ is of the form

$$Z(X;T) = \frac{P_1(X;T)}{(1-T)(1-qT)}$$

and the Hasse-Weil zeta-function $\zeta(X;s) = Z(X;q^{-s})$ has a pole of order 1 at $s = 1$ with residue

$$\mathrm{Res}_{s=1}\, \zeta(X;q^{-s}) = \frac{P_1(X;1/q)}{(1-1/q)} = \pm\frac{|\mathrm{Pic}^0(X)|/q}{(q-1)/q} = \pm\frac{|\mathrm{Pic}^0(X)|}{q-1}.$$

Observe that $|\mathrm{Pic}^0(X)|$ is nothing but the class number of the function field, $k(X)$ , of $X$ over $k$ . Therefore, the residue formula may be regarded as an analogue of the class number formula for the Dedekind zeta-function of a number field.

For higher dimensional cases, Tate [41] has formulated a conjecture linking the order $\rho_i$ of poles of $\zeta(X;s)$ at $s = i$ with the cohomology classes of algebraic cycles of codimension $i$ of $X$ . Further, Tate has laid the groundwork on the evaluation of the special values of $\zeta(X;s)$ at $s = i$ , $0 \le i \le 2d$ , relating them with various cohomological invariants of $X$ . Indeed, for $d = 2$ , Tate [43] has proved the conjecture using the $\ell$-adic étale cohomology for $\ell$ prime $\neq p$ , and Milne [24] has filled in the p-part using the formalism of crystalline and flat cohomologies. The higher dimensional cases $d > 2$ have subsequently been worked out by Lichtenbaum [17] and Milne [26].

In this article, we shall confine ourselves to the case where $X$ is a surface $(d = 2)$ defined over a finite field $k = \mathbb{F}_q$ , and purport to give a survey on the determination of the invariants of $X$ , e.g., the order $\rho_1$ of poles of $\zeta(X;s)$ at $s = 1$ , the Picard number , $\rho(X)$ , the order of the Brauer group, $|Br(X)|$ , and the discriminant of the Néron-Severi group, $\det NS(X)$ , among others, analysing the values of $P_2(X;q^{-s})$ at $s = 0$ and $1$ . In some cases, we can determine the values $|Br(X)|$ , and $\det NS(X)$ explicitly, incorporating results by Milne [23] and Yui [47] for product surfaces, Shioda [34] for Kummer surfaces, and Shioda [37] and Suwa and Yui [40] for Fermat surfaces (Fermat 2-motives).

The principal theme of this paper is that we first fix a finite field $k = \mathbb{F}_q$ of definition for an algebraic surface $X$ , and then deduce the arithmetical invariants of $X$ with respect to $k$ , from the congruence zeta-function of $X$ by evaluating its special values at $s = 0$ and $1$ . This is realized for abelian surfaces, K3 surfaces (Kummer surfaces) and Fermat surfaces (Fermat 2-motives).

Now we describe briefly the contents of the paper.

In Section 1 , we recall the formalism and some results on the crystalline cohomology groups and on the de Rham-Witt complex and Hodge-Witt cohomology groups for surfaces, which are relevant to the subsequent discussions.

In Section 2, the Tate conjecture and its validity for surfaces are reviewed, and the Artin-Tate formula is discussed in detail.

In Section 3, the invariant $\rho_1 = \rho(X)$ , the values of $P_2(X;q^{-s})$ at $s = 0$ and $1$ , and in some cases $|Br(X)|$ and $\det NS(X)$ are explicitly calculated relative to $k = \mathbb{F}_q$ , for abelian surfaces $X$ . The Honda-Tate theory for abelian varieties over finite fields plays the instrumental role in this analysis.

In Section 4 , K3 surfaces over $k = \mathbb{F}_q$ are discussed. Since it is rather hard to calculate the invariants for K3 surfaces in general, the actual computations are carried out for Kummer surfaces, making use of the results in Section 3 .

Finally in Section 5, the invariants of Fermat surfaces $X$ over $k = \mathbb{F}_q$ are evaluated by passing to Fermat 2-motives. The Jacobi sums, which are the eigenvalues of the geometric Frobenius $\Phi$ , play the essential role in this analysis. An algorithm for calculating the Picard number $\rho(X_{\bar{k}})$ is obtained. The values of $P_2(X;q^{-s})$ at $s = 0$ and $1$ are evaluated by passing to Fermat 2-motives and then calculating norms of algebraic integers associated to Jacobi sums.

**Acknowledgements**

N. Suwa was a visiting researcher at the Department of Mathematics and Statistics, Queen's University for two months, July and August 1988, partially supported by the Natural Sciences and Engineering Research Council of Canada through N. Yui's Operating Grants. He is grateful to Queen's University for the hospitality shown to him during his stay.

N. Yui is indebted to Harvey Cohn, David Chudnovsky and Gregory Chudnovsky for their interest in this work and encouragement.

We are grateful to D. Zagier for his permission to publish here the computational results on the norms of algebraic integers associated to Jacobi sums, and to T. Shioda for his constant encouragement.

**Notations.**

Throughout the paper, we employ the following notations:

$X$ : a smooth projective variety of dimension $d$ , mostly $d = 2$ ,

$p$ : a prime number,

$k$ : a perfect field of characteristic $p > 0$ , mostly a finite field $\mathbb{F}_q$ with $q = p^a$ elements,

$\bar{k}$ : the algebraic closure of $k$ ,

$\Gamma = \mathrm{Gal}(\bar{k}/k)$ : the Galois group of $\bar{k}$ over $k$ ,

$\ell$ : a prime number, mostly different from $p = \mathrm{char}(k)$ ,

$\mathbb{Q}_\ell$ : the field of $\ell$-adic rational numbers ,

$\mathbb{Z}_\ell$ : the ring of $\ell$-adic integers ,

$|\ |_\ell$ : the $\ell$-adic absolute value normalized so that $|\ell|_\ell^{-1} = \ell$ ,

$W = W(k)$ : the ring of infinite Witt vectors over $k$ ,

$K$ : the field of quotients of $W$ ,

$\Phi$ : the geometric Frobenius of $X$ relative to $k$ ,

$\phi$ : the arithmetic Frobenius of $X$ relative to $k$ ,

$H^i(X_{\bar{k}}, \mathbb{Q}_\ell)$ : the i-th $\ell$-adic étale cohomology group of $X$ ,

$H^i(X/W)_K$ : the i-th crystalline cohomology group of $X$ ,

$W\Omega^{\cdot}_X$ : the de Rham-Witt complex on $X$ ,

$W\Omega^{\cdot}_{X,\log}$ : the logarithmic Hodge-Witt sheaf on $X$ ,

$h^{ij}(X) = \dim_k H^j(X, \Omega^i_X)$ : the $(i,j)$-th Hodge number of $X$ ,

$p_g(X) = \dim_k H^2(X, \mathcal{O}_X) = h^{20}(X)$ : the geometric genus of $X$ ,

$B_i(X) = \dim_{\mathbb{Q}_\ell} H^i(X_{\bar{k}}, \mathbb{Q}_\ell) = \dim_K H^i(X/W)_K$ : the i-th Betti number of $X$ ,

$\mathrm{Pic}_{X/k}$ : the Picard scheme of $X$ ,

$NS(X)$ : the Néron-Severi group of $X$ ,

If $S$ is a set (group), $|S|$ denotes its cardinality (order),

$\mu_m$ $(m \in \mathbb{Z}, m > 0)$ : the group scheme of $m^{\text{th}}$ roots of unity.

## 1. Crystalline cohomology and de Rham-Witt complex

In this section, we shall recall the formalism and some results on crystalline cohomology and de Rham-Witt complex, confining ourselves to surfaces. On this subject, survey lectures by Illusie [10], [11], [12] should be helpful.

Let $k$ be a perfect field of characteristic $p > 0$, and let $X$ be a smooth projective surface defined over $k$.

**1.1.** The hypercohomology group of the de Rham-Witt complex $W\Omega^{\cdot}_X$ on $X$ is isomorphic to the crystalline cohomology group $H^{\cdot}(X/W)$ of $X$ (Illusie [13, Ch. II.2]). Furthermore, the decreasing filtration $(W\Omega^{\geq i}_X)$ of the complex $W\Omega^{\cdot}_X$ gives rise to a spectral sequence, called the *slope spectral sequence of* $X$, $E_1^{ij} = H^j(X, W\Omega^i_X) \Longrightarrow H^{i+j}(X/W)$.

**1.2.** The slope spectral sequence

$$E_1^{ij} = H^j(X, W\Omega^i_X) \Longrightarrow H^{i+j}(X/W) ,$$

degerates at $E_1$ modulo torsion. This yields an isomorphism

$$H^n(X/W)_K \simeq \bigoplus_{i+j=n} H^j(X, W\Omega^i_X)_K ;$$

more precisely, we have isomorphisms

$$H^n(X/W)_K^{[i,i+1[} \simeq H^{n-i}(X, W\Omega^i_X)_K . \quad ([13, \text{Ch. II.3}].)$$

This may be tabulated in the following diagram.

| | [0,1[ | [1,2[ | [2] |
|---|---|---|---|
| $H^4(X/W)_K$ | | | $H^2(X, W\Omega^2_X)_K$ |
| $H^3(X/W)_K$ | | $H^2(X, W\Omega^1_X)_K$ | $H^1(X, W\Omega^2_X)_K$ |
| $H^2(X/W)_K$ | $H^2(X, W\Omega^2_X)_K$ | $H^1(X, W\Omega^1_X)_K$ | $H^0(X, W\Omega^2_X)_K$ |
| $H^1(X/W)_K$ | $H^1(X, W\Omega^1_X)_K$ | $H^0(X, W\Omega^1_X)_K$ | |
| $H^0(X/W)_K$ | $H^0(X, W\mathcal{O}_X)_K$ | | |

**1.3.** The slope spectral sequence

$$E_1^{ij} = H^j(X, W\Omega_X^i) \Longrightarrow H^{i+j}(X/W)$$

degenerates at $E_2$ ([13, Ch. II. Cor. 3.13]) .

The slope spectral sequence of X is illustrated as follows:

$$\begin{array}{|lll} H^2(X,W\mathcal{O}_X) \xrightarrow{\ d\ } & H^2(X,W\Omega_X^1) & H^2(X,W\Omega_X^2) \\ H^1(X,W\mathcal{O}_X) & H^1(X,W\Omega_X^1) & H^1(X,W\Omega_X^2) \\ H^0(X,W\mathcal{O}_X) & H^0(X,W\Omega_X^1) & H^0(X,W\Omega_X^2) \\ \hline \end{array}$$

Furthermore, the following assertions hold.

**1.4. Theorem.** (Nygaard [27, Th. 2.4].)

(1) The differential $d : H^j(X, W\Omega_X^i) \longrightarrow H^j(X, W\Omega_X^{i+1})$ is zero except for $(i, j) = (0,2)$ .

(2) The following conditions are all equivalent.

(i) $H^2(X, W\mathcal{O}_X)$ is of finite type over $W$ .

(ii) The differential $d : H^2(X, W\mathcal{O}_X) \longrightarrow H^2(X, W\Omega_X^1)$ is zero.

(iii). The slope spectral sequence

$$E_1^{ij} = H^j(X, W\Omega_X^i) \Longrightarrow H^{i+j}(X/W)$$

degenerates at $E_1$ .

**1.4.1. Definition.** X is said to be *of Hodge-Witt type* if X satisfies one of the conditions (i) , (ii) or (iii) of (1.4) .

**1.5.** Some properties of $H^j(X, W\Omega_X^i)$ are listed in the following:

(1) $H^0(X, W\mathcal{O}_X) \simeq H^0(X/W) \simeq W$ , $H^2(X, W\Omega_X^2) \simeq H^4(X/W) \simeq W$ .

(2) $H^j(X, W\Omega_X^i)$ are W-modules of finite type except for $(i, j) = (0,2)$ and $(1,2)$ .

$H^1(X, W\mathcal{O}_X)$ , $H^0(X, W\Omega_X^1)$ and $H^0(X, W\Omega_X^2)$ are all torsion-free.

(3) $H^1(X, W\mathcal{O}_X)$ is isomorphic to the Cartier module of the formal Picard group $(\mathrm{Pic}^0_{X/k,\mathrm{red}})\hat{}$ . Moreover, there exists an exact sequence of W-modules

$$0 \longrightarrow H^0(X, W\Omega^1_X) \longrightarrow H^1(X/W) \longrightarrow H^1(X, W\mathcal{O}_X) \longrightarrow 0 ,$$

which corresponds to the exact sequence of smooth formal groups over $k$ :

$$0 \longrightarrow ({}_{p^\infty}\mathrm{Pic}_{X/k,\mathrm{red}})^0 \longrightarrow {}_{p^\infty}\mathrm{Pic}_{X/k,\mathrm{red}} \longrightarrow ({}_{p^\infty}\mathrm{Pic}_{X/k,\mathrm{red}})^{et} \longrightarrow 0 .$$

(4) $H^2(X, W\mathcal{O}_X)$ is isomorphic to the Cartier module of the formal Brauer group $\hat{Br}_{X/k} = H^2(X, \hat{\mathbb{G}}_{m,X})$ when $H^2(X, \hat{\mathbb{G}}_{m,X})$ is pro-representable. In this case,

$$\dim H^2(X, W\mathcal{O}_X)_K = \dim H^2(X/W)_K^{[0,1[}$$

$$= \text{the height of the } p\text{-divisible part of } \hat{Br}_{X/k} .$$

([13, Ch. II.2]).

**1.5.1. Remark.** $H^2(X, \hat{\mathbb{G}}_{m,X})$ is pro-representable by a smooth formal group over $k$ if and only if the Picard scheme $\mathrm{Pic}_{X/k}$ is smooth over $k$

1.6. Put

$$T^{ij} = \begin{cases} \mathrm{length}_{W((V))} H^2(X, W\mathcal{O}_X) \otimes_{W[[V]]} W((V)) & (i,j) = (0,2) \\ 0 & (i,j) \neq (0,2) \end{cases}$$

and

$$h_W^{i,n-i}(X) = \sum_{\lambda \in [i-1,i[} m_\lambda(\lambda - i + 1) + \sum_{\lambda \in [i,i+1[} m_\lambda(i+1-\lambda)$$

$$+ T^{i,n-i} - 2T^{i-1,n-i+1} + T^{i-2,n-i+2}$$

where $m_\lambda$ is the multiplicity of a slope $\lambda$ in $H^n(X/W)_K$. We call $h_W^{ij}(X)$ the $(i,j)$-th *Hodge-Witt number* of $X$ (Ekedhal [6]).

Recall that the $(i,j)$-th *Hodge number*, $h^{ij}(X)$, is defined by

$$h^{ij}(X) = \dim_k H^j(X, \Omega^i_X) .$$

The Hodge-Witt numbers $h_W^{ij}(X)$ and the Hodge numbers $h^{ij}(X)$ are related, and indeed we have

$$h_W^{01}(X) = h_W^{10}(X) = h_W^{21}(X) = h_W^{12}(X) = \dim \mathrm{Alb}_{X/k} = B_1(X)/2 ,$$

and

$$h_W^{02}(X) = h_W^{20}(X) = \chi(\mathcal{O}_X) - 1 + B_1(X)/2 = h^{02}(X) - \delta(X)$$

where

$$\delta(X) = h^{10}(X) - \dim \mathrm{Alb}_{X/k} \quad \text{(the "defect of smoothness" of } \mathrm{Pic}_{X/k}\text{)}$$

(Crew [4].)

**1.6.1. Remarks.** (1) $h_W^{10}(X) = 0$ if and only if $\mathrm{Alb}_{X/k} = 0$ .

(2) $h_W^{20}(X) = 0$ if and only if $H^2(X, W\mathcal{O}_X)$ is of V-torsion.

(Suwa [39 , Prop. 4].)

**1.7.** Let $W\Omega^{\cdot}_{X,\log}$ denote the logarithmic Hodge-Witt sheaf on $X$ defined by Illusie [13, Ch. I.5]. We put

$$H^j(X, \mathbb{Z}_p(i)) = H^{j-i}(X, W\Omega^i_{X,\log})$$

and

$$H^j(X, \mathbb{Q}_p(i)) = \quad H^j(X, \mathbb{Z}_p(i)) \otimes_{\mathbb{Z}_p} \mathbb{Q}_p .$$

**1.7.1. Remarks.** (1) $H^j(X, \mathbb{Z}_p(0))$ is indeed the p-adic étale cohomology group $H^j(X, \mathbb{Z}_p) = \varprojlim H^j(X_{et}, \mathbb{Z}/p^n)$ .

(2) $H^j(X, \mathbb{Z}_p(1))$ is isomorphic to the flat cohomology group $\varprojlim H^j(X_{fl}, \mu_{p^n})$ .

(loc. cit. Ch. II.5 .)

**1.8.** In this section we assume that $k$ is algebraically closed. The sequence of pro-sheaves on $X_{et}$

$$0 \longrightarrow W\Omega^i_{X,\log} \longrightarrow W\Omega^i_X \xrightarrow{F-1} W\Omega^i_X \longrightarrow 0$$

induces the exact sequences

$$0 \longrightarrow H^j(X, \mathbb{Z}_p(i)) \longrightarrow H^{j-i}(X, W\Omega^i_X) \xrightarrow{F-1} H^{j-i}(X, W\Omega^i_X) \longrightarrow 0 .$$

We have an isomorphism

$$H^j(X, \mathbb{Q}_p(i)) = H^{j-i}(X, W\Omega^i_X)_K^{F=1} = H^j(X/W)_K^{F=p^i} .$$

(Illusie-Raynaud [14, Ch. IV.3].)

1.8.1. Some properties of $H^j(X, \mathbb{Z}_p(i))$ are listed in the following.

(1) $H^j(X, \mathbb{Z}_p(i))$ are $\mathbb{Z}_p$-modules of finite type except for $(i,j) = (1,3)$.

$H^1(X, \mathbb{Z}_p)$ , $H^1(X, \mathbb{Z}_p(1))$ and $H^2(X, \mathbb{Z}_p(2))$ are all torsion-free.

(2) There exists a canonical exact sequence

$$0 \longrightarrow \underline{U}^3(X, \mathbb{Z}_p(1))(k) \longrightarrow H^3(X, \mathbb{Z}_p(1)) \longrightarrow \underline{D}^3(X, \mathbb{Z}_p(1))(k) \longrightarrow 0 ,$$

where $\underline{U}^3(X, \mathbb{Z}_p(1))$ is a connected unipotent quasi-algebraic k-group and $\underline{D}^3(X, \mathbb{Z}_p(1))$ is a pro-étale k-group of finite type over $\mathbb{Z}_p$. Furthermore, $\dim \underline{U}^3(X, \mathbb{Z}_p(1)) = T^{02}$. (Milne [25] and Illusie-Raynaud [14, Ch. IV. Th.3.3]).

We tabulate the cohomology groups $H^j(X, \mathbb{Z}_p(i))$ in the following diagram:

| | | |
|---|---|---|
| | | $H^4(X, \mathbb{Z}_p(2)) = \mathbb{Z}_p$ |
| | $H^3(X, \mathbb{Z}_p(1))$ | $H^3(X, \mathbb{Z}_p(2))$ |
| $H^2(X, \mathbb{Z}_p)$ | $H^2(X, \mathbb{Z}_p(1))$ | $H^2(X, \mathbb{Z}_p(2))$ |
| $H^1(X, \mathbb{Z}_p)$ | $H^1(X, \mathbb{Z}_p(1))$ | |
| $H^0(X, \mathbb{Z}_p) = \mathbb{Z}_p$ | | |

**1.9.** $H^n(X/W)/\text{tors}$ has the structure of an F-crystal. The Newton polygon and the Hodge polygon of $H^n(X/W)$ are defined in terms of semi-linear algebras. It is known that the Newton polygon lies over or on the Hodge polygon. (Cf. Katz [15, Ch. I].)

There is a deep connection between the (abstract) Hodge numbers and the geometric Hodge numbers.

**1.9.1.** If $H^n(X/W)$ is torsion-free and if the Hodge spectral sequence $E_1^{ij} = H^j(X,\Omega_X^i) \Longrightarrow H_{DR}^{i+j}(X/k)$ degenerates at $E_1$, then the (abstract) Hodge numbers of $H^n(X/W)$ are indeed given by the geometric Hodge numbers $h^{ij}(X)$ (Mazur [21] and Nygaard [28].)

The Hodge polygon of $H^n(X/W)$ is defined to be the graph of the real-valued continuous piece-wise linear function on the interval $[0,B_n(X)]$ with the initial point $(0,0)$ and with slopes $i$ of multiplicities $h^{ij}(X)$, respectively.

The surfaces that we shall consider in the subsequent sections all satisfy the above conditions. Hence we are able to determine their Hodge numbers and the Hodge polygons explicitly.

**1.9.2.** If $k$ is a finite field $\mathbb{F}_q$, the Newton polygon of $H^n(X/W)_K$ is determined by the characteristic polynomial $P_n(X;T) = \det(1-\Phi T; H^n(X/W)_K) \in \mathbb{Z}[T]$.

We fix a p-adic valuation $\nu$ of $\mathbb{Q}_p$ normalized so that $\nu(q) = 1$. Let $\alpha_1,\alpha_2,\ldots,\alpha_{B_n(X)}$ be the reciprocal roots of $P_n(X;T) = 0$ in the algebraic closure $\overline{\mathbb{Q}}_p$ of $\mathbb{Q}_p$. Then the slopes of $P_n(X;T)$ are

$$0 \le \nu(\alpha_1) \le \nu(\alpha_2) \le \ldots \le \nu(\alpha_{B_n(X)})$$

arranged in non-decreasing order (changing subindecies if necessary). The Newton polygon of $P_n(X;T)$ is defined to be the graph of the real-valued continuous piece-wise linear function on the interval $[0,B_n(X)]$ with the initial point $(0,0)$ and with the derivative $\nu(\alpha_j)$ in the interval $]j-1,j[$. Therefore, the Newton polygon of $H^n(X/W)_K$ coincides with the Newton polygon of $P_n(X;T)$. (Manin [19, Ch. 4].)

## 2. The Tate conjecture and the Artin-Tate formula

Let $X$ be a smooth projective variety over a finite field $k = \mathbb{F}_q$ of characteristic $p > 0$ and let $\Gamma = \mathrm{Gal}(\bar{k}/k)$ be the Galois group of $\bar{k}$ over $k$.

**2.1.** The Tate conjecture asserts that

*$H^{2i}(X_{\bar{k}}, \mathbb{Q}_\ell(i))$ ($\ell$ prime $\neq p$) is spanned by the cohomology classes of algebraic cycles of $X$ of codimension $i$ defined over $k$.*

This is equivalently formulated as follows:

*The multiplicity, $\rho_i$, of $q^i$ as a reciprocal root of the polynomial $P_{2i}(X;T) = \det(1-\Phi T; H^{2i}(X_{\bar{k}}, \mathbb{Q}_\ell))$ coincides with the dimension of the subspace of $H^{2i}(X_{\bar{k}}, \mathbb{Q}_\ell(i))$ spanned by the cohomology classes of algebraic cycles of codimension $i$ defined over $k$.*

(Tate [41].)

**2.2.** We now describe the Tate conjecture more precisely in the case of codimension 1. Let $NS(X_{\bar{k}}) = \mathrm{Pic}(X_{\bar{k}})/\mathrm{Pic}^0(X_{\bar{k}})$ denote the group of the divisors on $X_{\bar{k}}$ modulo the algebraic equivalence and let $NS(X)$ denote the subgroup of $NS(X_{\bar{k}})$ generated by the divisors on $X$, i.e., the divisors on $X_{\bar{k}}$ defined over $k$. $NS(X_{\bar{k}})$ (resp. $NS(X)$) is called the *Néron-Severi group of* $X_{\bar{k}}$ (resp. $X$). Since $k$ is finite, $H^1(\bar{k}/k, \mathrm{Pic}^0(X_{\bar{k}})) = 0$ (Lang's theorem). Therefore we have $NS(X) = NS(X_{\bar{k}})^\Gamma$. It is known that $NS(X_{\bar{k}})$ is a $\mathbb{Z}$-module of finite type (the theorem of Néron-Severi). We denote by $\rho(X_{\bar{k}})$ (resp. $\rho(X)$) the rank of $NS(X_{\bar{k}})$ (resp. $NS(X)$) and call it the *Picard number* of $X_{\bar{k}}$ (resp. $X$).

For a prime $\ell \neq p$, the Kummer sequences

$$0 \longrightarrow \mu_{\ell^n} \longrightarrow \mathbb{G}_m \xrightarrow{\ \ell^n\ } \mathbb{G}_m \longrightarrow 0$$

induce an exact sequence

$$0 \longrightarrow NS(X_{\bar{k}})\otimes_{\mathbb{Z}} \mathbb{Z}_\ell \longrightarrow H^2(X_{\bar{k}},\mathbb{Z}_\ell(1)) \longrightarrow T_\ell H^2(X_{\bar{k}},\mathbb{G}_m) \longrightarrow 0 .$$

Hence we can reformulate the Tate conjecture in the case of codimension 1 as follows:

*The multiplicity* $\rho_1$ *of* $q$ *as a reciprocical root of the polynomial* $P_2(X;T) = \det(1-\Phi T; H^2(X_{\bar{k}},\mathbb{Q}_\ell))$ *coincides with the Picard number* $\rho(X)$ *of* $X$ , *i.e.*, $\rho_1 = \rho(X)$ .

Note also that the above exact sequence induces an isomorphism

$$NS(X_{\bar{k}})_{\ell\text{-tors}} \simeq H^2(X_{\bar{k}},\mathbb{Z}_\ell(1))_{\text{tors}} .$$

**2.3. Remarks.** For $\ell = p$ , the Kummer sequence gives rise to deep relations between the Picard number and the second crystalline cohomology group of $X$ . In fact, the Kummer sequences

$$0 \longrightarrow \mu_{p^n} \longrightarrow \mathbb{G}_m \xrightarrow{p^n} \mathbb{G}_m \longrightarrow 0$$

induce an exact sequence

$$0 \longrightarrow NS(X_{\bar{k}})\otimes_{\mathbb{Z}} \mathbb{Z}_p \longrightarrow H^2(X_{\bar{k}},\mathbb{Z}_p(1)) \longrightarrow T_p H^2(X_{\bar{k}},\mathbb{G}_m) \longrightarrow 0 .$$

From this we can deduce the following facts:

(1) $NS(X_{\bar{k}})$ is isomorphic to $H^2(X_{\bar{k}},\mathbb{Z}_p(1))$ .

(2) (The Igusa-Artin-Mazur inequality)

$$\rho(X_{\bar{k}}) \le \operatorname{rk} H^2(X_{\bar{k}},\mathbb{Z}_p(1)) = \dim_K H^2(X/W)_K^{[1]} .$$

Further, if $H^2(X,\hat{\mathbb{G}}_{m,X}) = \widehat{Br}_{X/k}$ is pro-representable by a smooth formal group, let $h$ denote the height of the p-divisible part of $\widehat{Br}_{X/k}$ . Then we have

$$\dim_K H^2(X/W)_K^{[1]} = B_2(X) - 2h$$

(Artin-Mazur [3] and Illusie [13 , Ch. II.5]).

Henceforth, the Tate conjecture means simply the Tate conjecture of codimension of 1 for surfaces.

**2.4.** We first list some elementary facts concerning the Tate conjecture.

(1) Let $X \longrightarrow Y$ be a dominant $k$-rational map of smooth projective surfaces over $k$. If the Tate conjecture holds for $X$, then so does for $Y$.

(2) Let $X$ be a smooth proejctive surface over $k$. If $\rho(X_{\bar{k}}) = B_2(X_{\bar{k}})$, then the Tate conjecture holds for $X$.

In particular, the Tate conjecture holds for any unirational surfaces (Shioda [33]).

**2.5.** The validity of the Tate conjecture has been established in the following cases:

(1) Abelian surfaces (or Abelian varieties) (Tate [42]),

(2) Products of two curves (Tate [42]),

(3) Fermat surfaces or Delsarte surfaces (Tate [41], Shioda and Katsura [38], Shioda [36]),

(4) K3 surfaces of finite height (Nygaard [29], Nygaard and Ogus [30]).

(5) Elliptic K3 surfaces (Artin and Swinnerton-Dyer [2]).

There are interrelations among these surfaces on the validity of the Tate conjecture.

(2) follows from (1), due to the formula

$$NS(X_{\bar{k}} \times Y_{\bar{k}}) = \mathbb{Z}^2 \oplus \mathrm{Hom}_{\bar{k}\text{-gr}}(\mathrm{Pic}^0_X, \mathrm{Pic}^0_Y)$$

where $X$ and $Y$ denote smooth projective curves over $k$. Moreover (3) follows from (2) because there exists a dominant rational map from the product of two curves to a Fermat surface or to a Delsarte surface (the inductive structure of Fermat surfaces or of Delsarte surfaces) (Shioda [36]).

(4) and (5) exhaust almost all cases of K3 surfaces. For the remaining cases, the validity of the Tate conjecture follows from that of the Artin-Mazur conjecture: If $X$ is supersingular (cf. (4.3.1)), then $\rho(X_{\bar{k}}) = B_2(X) = 22$.

**2.6. Remark.** Tate [41] has formulated a more general conjecture for varieties defined over finitely generated fields over prime fields. Faltings [7] has succeeded in proving the conjecture of codimension 1 for abelian varieties over number fields. It should be pointed out that the validity of the conjecture for products of curves, Fermat surfaces and K3 surfaces over number fields is in fact a consequence of the theorem of Faltings on abelian varieties.

**2.7. Proposition.** Let $X$ be a smooth projective surface over $k = \mathbb{F}_q$. Assume the validity of the Tate conjecture for $X$. Then the following assertions hold for $X$.

(1) $B_2(X) - \rho(X_{\bar{k}}) \equiv 0 \pmod{2}$.

(2) If $H^2(X/W)_K$ has pure slope $1$, then $\rho(X_{\bar{k}}) = B_2(X)$.

**Proof.** Let $S = \{ \alpha_i | 1 \le i \le B_2(X) \}$ be the set of reciprocal roots of $P_2(X;T) = 0$.

(1) We may assume that $k$ is sufficiently large so that $\rho(X_{\bar{k}}) = \rho(X)$. By Poincaré duality, the map $\alpha_i \longrightarrow q^2/\alpha_i$ is a permutation of order $2$ on $S$. Thus

$$B_2(X) - |\{ \alpha_i \mid \alpha_i = q^2/\alpha_i \}| \equiv 0 \pmod{2}.$$

Now the validity of the Tate conjecture:

$$\rho(X_{\bar{k}}) = \rho(X) = |\{ \alpha_i \mid \alpha_i = q \}|$$

implies that there exist no $\alpha_i \in S$ such that $\alpha_i = -q$. Hence

$$|\{ \alpha_i \mid \alpha_i = q^2/\alpha_i \}| = \rho(X).$$

(2) The hypothesis and conventions of (1) remain in force. The reciprocal roots $\alpha_i$ satisfy the Weil-Riemann hypothesis: $\alpha_i/q$ has the absolute value $1$ in any embedding into $\mathbb{C}$ for all $i = 1,\ldots,B_2(X)$. For $\ell$ prime $\neq p$, $\alpha_i/q$ are $\ell$-adic units for all $i$. For $\ell = p$, the hypothesis that $H^2(X/W)_K$ has pure slope $1$ assures that $\alpha_i/q$ are p-adic units for all $i$. Therefore, $\alpha_i/q$ are roots of unity for all $i$. Now the assumption $\rho(X_{\bar{k}}) = \rho(X)$ implies that $\alpha_i = q$ for all $i$. Finally the validity of the Tate conjecture asserts that $\rho(X) = B_2(X)$.

**2.7.1. Remark.** The statement (2) of Proposition (2.7) is the Artin-Mazur conjecture for $X$ (Artin and Mazur [3]). Therefore, the validity of the Tate conjecture implies that of the Artin-Mazur conjecture for surfaces over finite fields.

Now we turn our discussion to the evaluation of the special values of the congruence zeta-function $\zeta(X;q^{-s})$ at $s = 0$ and $s = 1$, concentrating mostly on the polynomial $P_2(X;T) = \det(1-\Phi T \,;\, H^2(X_{\bar{k}},\mathbb{Q}_\ell^1)) = \det(1-\Phi T \,;\, H^2(X/W)_K)$.

**2.8. Theorem.** *Let* $\{\alpha\}$ *be the set of reciprocal roots of* $P_2(X;T) = 0$ . *Then we have the following formula:*

$$P_2(X;1) = \prod_{\alpha} (1 - \alpha/q)$$

$$= \frac{|H^2(X,\mathbb{Q}/\mathbb{Z})|}{|H^1(X_{\bar{k}},\mathbb{Q}/\mathbb{Z})^{\Gamma}_{\text{cotors}}|\,|H^2(X_{\bar{k}},\mathbb{Q}/\mathbb{Z})^{\Gamma}_{\text{cotors}}|} .$$

**2.8.1.** In what follow, we shall give an outline of proof of Theorem (2.8).

Pick a prime $\ell \neq p$ . Then it is well known (cf. Milne [24]) that $\Phi = \phi^{-1}$ on $H^{\cdot}(X_{\bar{k}},\mathbb{Q}_\ell)$ . This implies that

$$\det(1 - \phi^{-1}T;\ H^2(X_{\bar{k}},\mathbb{Q}_\ell(r)) = \prod_{\alpha} (1 - \frac{\alpha}{q^r}\,T) .$$

Assume now that $r \neq 1$ . We substitute $T = 1$ to the above formula to get

$$\det(1 - \phi^{-1};\ H^2(X_{\bar{k}},\mathbb{Q}_\ell(r))) = P_2(X\,;\frac{1}{q^r}) .$$

Moreover,

$$|\det(1 - \phi^{-1};\ H^2(X_{\bar{k}},\ \mathbb{Q}_\ell(r)))|_\ell^{-1} = |(H^2(X_{\bar{k}},\mathbb{Z}_\ell(r))/\text{tors})_\Gamma| .$$

Furthermore, the following exact sequence, which is induced from the Hochschild-Serre spectral sequence,

$$0 \longrightarrow H^2(X_{\bar{k}},\mathbb{Z}_\ell(r))_\Gamma \longrightarrow H^3(X,\mathbb{Z}_\ell(r)) \longrightarrow H^3(X_{\bar{k}},\mathbb{Z}_\ell(r))^\Gamma \longrightarrow 0$$

gives rise to the identity

$$|(H^2(X_{\bar{k}},\mathbb{Z}_\ell(r))/\text{tors})_\Gamma|$$

$$= \frac{|H^3(X,\mathbb{Z}_\ell(r))|}{|H^2(X_{\bar{k}},\mathbb{Z}_\ell(r))^{\Gamma}_{\text{tors}}|\ |H^3(X_{\bar{k}},\mathbb{Z}_\ell(r))^{\Gamma}_{\text{tors}}|} .$$

Now let $r = 0$ . Then the Bockstein operator yields the following isomorphisms

$$H^i(X,\mathbb{Z}_\ell) \xleftarrow{\sim} H^{i-1}(X,\mathbb{Q}_\ell/\mathbb{Z}_\ell) \xleftarrow{\sim} H^{i-1}(X,\mathbb{Q}/\mathbb{Z})_{\ell\text{-tors}}$$

for $i > 1$, and

$$H^i(X_{\bar{k}},\mathbb{Z}_\ell)^\Gamma_{\text{tors}} \xleftarrow{\sim} H^{i-1}(X_{\bar{k}},\mathbb{Q}_\ell/\mathbb{Z}_\ell)^\Gamma_{\text{cotors}} \xleftarrow{\sim} H^{i-1}(X_{\bar{k}},\mathbb{Q}/\mathbb{Z})^\Gamma_{\ell\text{-cotors}}$$

for $i \geq 1$. Thus, we obtain

$$|P_2(X;1)|_\ell^{-1} = \frac{|H^2(X,\mathbb{Q}/\mathbb{Z})_{\ell\text{-tors}}|}{|H^1(X_{\bar{k}},\mathbb{Q}/\mathbb{Z})^\Gamma_{\ell\text{-cotors}}||H^2(X_{\bar{k}},\mathbb{Q}/\mathbb{Z})^\Gamma_{\ell\text{-cotors}}|} .$$

Now we consider the case $\ell = p$. We have

$$H^2(X_{\bar{k}},\mathbb{Q}_p) = [H^2(X_{\bar{k}}/W(\bar{k}))\otimes_{\mathbb{Z}} \mathbb{Q}]^{F=1}$$

and this implies that

$$\det(1 - \phi^{-1}T \; ; \; H^2(X_{\bar{k}},\mathbb{Q}_p)) = \prod_{\nu(\alpha)=0} (1 - \alpha T) .$$

Substituting $T = 1$ in the above formula, we get

$$\det(1 - \phi^{-1} \; ; \; H^2(X_{\bar{k}},\mathbb{Q}_p)) = \prod_{\nu(\alpha)=0} (1 - \alpha) .$$

Moreover,

$$|\det(1 - \phi^{-1} \; ; \; H^2(X_{\bar{k}},\mathbb{Q}_p))|_p^{-1} = |((H^2(X_{\bar{k}},\mathbb{Z}_p)/\text{tors})_\Gamma| .$$

Observe that

$$|\prod_{\nu(\alpha)=0} (1 - \alpha)|_p^{-1} = |\prod_\alpha (1 - \alpha)|_p^{-1}$$

(as $1-\alpha$ is a p-adic unit if $\nu(\alpha) > 0$). Thus

$$|P_2(X;1)|_p^{-1} = |(H^2(X_{\bar{k}},\mathbb{Z}_p)/\text{tors})_\Gamma| .$$

By applying the same line of arguments as for the case $\ell \neq p$, we obtain

$$|P_2(X;1)|_p^{-1} = \frac{|H^2(X,\mathbb{Q}/\mathbb{Z})_{p\text{-tors}}|}{|H^1(X_{\bar{k}},\mathbb{Q}/\mathbb{Z})^\Gamma_{p\text{-cotors}}| \; |H^2(X_{\bar{k}},\mathbb{Q}/\mathbb{Z})^\Gamma_{p\text{-cotors}}|} .$$

Finally, putting together all the cases, we arrive at the required formula.

**2.9. Theorem (The Artin-Tate formula).** *Let* $\{\alpha\}$ *be the set of reciprocal roots of* $P_2(X;T) = 0$ . *Assume the validity of the Tate conjecture. Then we have the following formula:*

$$\left[\frac{P_2(X;T)}{(1-qT)^{\rho(X)}}\right]_{T=1/q} = \prod_{\alpha\neq q} (1 - \alpha/q)$$

$$= \pm \frac{|Br(X)||\det NS(X)|}{|NS(X)_{tors}|^2 \; q^{h_W^{20}(X)}} .$$

**2.9.1.** We shall briefly recall the method of Tate [41] and Milne [23,24] for the evaluation of $P_2(X;q^{-s})$ at $s = 1$ .

Let $M$ be an abelian group and let $\gamma : M \longrightarrow M$ be an endomorphism. The quantity

$$z(\gamma) = \frac{|Ker(\gamma)|}{|Coker(\gamma)|}$$

is defined provided that both $|Ker(\gamma)|$ and $|Coker(\gamma)|$ are finite.

Pick any prime $\ell$ (including $\ell = p$) . We have a commutative diagram

$$\begin{array}{ccc} H^2(X,\mathbb{Z}_\ell(1)) & \xrightarrow{\varepsilon_\ell} & H^3(X,\mathbb{Z}_\ell(1)) \\ \downarrow h_\ell & & \uparrow g_\ell \\ H^2(X_{\bar{k}},\mathbb{Z}_\ell(1)) & \xrightarrow{f_\ell} & H^2(X_{\bar{k}},\mathbb{Z}_\ell(1)) \end{array}$$

where the vertical arrows come from the Hochschild-Serre spectral sequence, $f_\ell$ is a composition $H^2(X_{\bar{k}},\mathbb{Z}_\ell(1))^\Gamma \longrightarrow H^2(X_{\bar{k}},\mathbb{Z}_\ell(1)) \longrightarrow H^2(X_{\bar{k}},\mathbb{Z}_\ell(1))_\Gamma$ and $\varepsilon_\ell$ is defined by cup product with $1 \in H^1(k,\mathbb{Z}_\ell) = \mathbb{Z}_\ell$ . (Milne [24]; Tate [43].)

Pick a prime $\ell \neq p$ . The validity of the Tate conjecture assures that $z(f_\ell) = |Ker(f_\ell)|/|Coker(f_\ell)|$ is well-defined, and furthermore,

$$z(f_\ell) = |\prod_{\alpha\neq q} (1 - \frac{\alpha}{q})|_\ell .$$

We also have

$$z(\varepsilon_\ell) = z(g_\ell)\, z(f_\ell)\, z(h_\ell) .$$

Moreover,

$$z(\varepsilon_\ell) = \frac{1}{\det\varepsilon_\ell} \cdot \frac{|H^2(X,\mathbb{Z}_\ell(1))_{tors}|}{|H^3(X,\mathbb{Z}_\ell(1))_{tors}|} ,$$

$$z(h_\ell) = |H^1(X_{\bar{k}},\mathbb{Z}_\ell(1))_\Gamma| ,$$

and

$$z(g_\ell) = 1/\ |H^3(X_{\bar{k}},\mathbb{Z}_\ell(1))^\Gamma| = 1/|H^3(X_{\bar{k}},\mathbb{Z}_\ell(1))^\Gamma_{tors}|$$

$$= 1/|H^2(X_{\bar{k}},\mathbb{Z}_\ell(1))^\Gamma_{tors}|$$

where the last equality is a consequence of the Poincaré duality.

Putting the above informations together, we can calculate $z(f_\ell)$ : Indeed, we have

$$z(f_\ell)^{-1} = |\prod_{\alpha\neq q} (1 - \frac{\alpha}{q})|_\ell^{-1} = z(\varepsilon_\ell)^{-1}\ z(g_\ell)\ z(h_\ell)$$

$$= \det\ \varepsilon_\ell \cdot \frac{|H^3(X,\mathbb{Z}_\ell(1))_{tors}|}{|H^2(X,\mathbb{Z}_\ell(1))_{tors}|} \cdot \frac{|H^1(X_{\bar{k}},\mathbb{Z}_\ell(1))_\Gamma|}{|H^2(X_{\bar{k}},\mathbb{Z}_\ell(1))^\Gamma_{tors}|}$$

$$= \det\ \varepsilon_\ell \cdot \frac{|H^3(X,\mathbb{Z}_\ell(1)_{tors}|}{|H^2(X_{\bar{k}},\mathbb{Z}_\ell(1))^\Gamma_{tors}|^2} .$$

Moreover, with the validity of the Tate conjecture, $H^3(X,\mathbb{Z}_\ell(1))$ is finite so that $Br(X)_{\ell\text{-tors}} \simeq H^3(X,\mathbb{Z}_\ell(1))_{tors}$ and $\det\ \varepsilon_\ell = \det NS(X) \otimes_{\mathbb{Z}} \mathbb{Z}_\ell$ .

Now consider the case $\ell = p$ . We have

$$H^2(X_{\bar{k}},\mathbb{Q}_p(1)) = [H^2(X_{\bar{k}}/W(\bar{k})) \otimes_{\mathbb{Z}} \mathbb{Q}\ ]^{F=p}$$

and this implies that

$$\det(1-\phi^{-1}T;\ H^2(X_{\bar{k}},\mathbb{Q}_p(1)) = \prod_{\nu(\alpha)=1} (1 - \frac{\alpha}{q}T) .$$

With the validity of the Tate conjecture, we can evaluate $z(f_p)$ : In fact, we obtain

$$z(f_p) = |\prod_{\substack{\nu(\alpha)=1\\ \alpha\neq q}} (1 - \frac{\alpha}{q})|_p$$

$$= |\prod_{\alpha\neq q} (1 - \frac{\alpha}{q})|_p \; |\prod_{\nu(\alpha)<1} (1 - \frac{\alpha}{q})|_p^{-1} \; |\prod_{\nu(\alpha)>1} (1 - \frac{\alpha}{q})|_p^{-1}$$

Here

$$|\prod_{\nu(\alpha)<1} (1 - \frac{\alpha}{q})|_p^{-1} = q^{-h_W^{02}(X)+T^{02}}$$

and

$$|\prod_{\nu(\alpha)>1} (1 - \frac{\alpha}{q})|_p^{-1} = 1 .$$

Moreover,

$$z(\varepsilon_p) = \frac{1}{\det \varepsilon_p} \cdot \frac{|H^2(X,\mathbb{Z}_p(1))_{tors}|}{|H^3(X,\mathbb{Z}_p(1))_{tors}|} ,$$

$$z(h_p) = |H^1(X_{\bar{k}},\mathbb{Z}_p(1))_\Gamma| ,$$

and

$$z(g_p) = 1/|H^3(X_{\bar{k}},\mathbb{Z}_p(1))^\Gamma| = 1/|H^3(X_{\bar{k}},\mathbb{Z}_p(1))^\Gamma_{tors}|$$

$$=1/[\,|\underline{D}^3(X,\mathbb{Z}_p(1))(k)_{tors}|\;|\underline{U}^3(X,\mathbb{Z}_p(1))(k)|\,] .$$

Further, by Milne's flat duality [24] we have

$$|\underline{D}^3(X,\mathbb{Z}_p(1))(k)_{tors}| = |H^2(X_{\bar{k}},\mathbb{Z}_p(1))^\Gamma_{tors}|$$

and

$$|\underline{U}^3(X,\mathbb{Z}_p(1))(k)| = q^{T^{02}} .$$

Combining the above quantities altogether, we obtain

$$|\prod_{\alpha\neq q} (1 - \frac{\alpha}{q})|_p^{-1} = \frac{|H^3(X,\mathbb{Z}_p(1))_{tors}|}{|H^2(X_{\bar{k}},\mathbb{Z}_p(1))^\Gamma_{tors}|^2} \cdot \frac{1}{q^{h_W^{02}(X)}} .$$

Again with the validity of the Tate conjecture, $H^3(X,\mathbb{Z}_p(1))$ is finite, so that $Br(X)_{p\text{-tors}} \simeq H^3(X,\mathbb{Z}_p(1))_{tors}$ and $\det \varepsilon_p = \det NS(X) \otimes_{\mathbb{Z}} \mathbb{Z}_p$ .

Finally, putting together all primes, we obtain the Artin-Tate formula.

**2.10. Remark.** Let $\ell$ be a prime ($\ell = p$ is included). As we have seen above, there exist isomorphisms of cohomology groups

$$(r = 0) \qquad H^{i+1}(X, \mathbb{Z}_\ell) \xleftarrow{\sim} H^i(X, \mathbb{Q}/\mathbb{Z})_{\ell\text{-tors}} \qquad (i \geq 1),$$

and

$$(r = 1) \qquad H^{i+1}(X, \mathbb{Z}_\ell(1)) \xleftarrow{\sim} H^i(X, \mathbb{G}_m)_{\ell\text{-tors}} \qquad (i \geq 2)$$

For $r = 2$, there are isomorphisms

$$H^2(X, \mathbb{Z}_\ell(2)) \xleftarrow{\sim} H^0(X_{zar}, \mathcal{K}_2)_{\ell\text{-tors}}$$

and

$$H^3(X, \mathbb{Z}_\ell(2)) \xleftarrow{\sim} H^1(X_{zar}, \mathcal{K}_2)_{\ell\text{-tors}}$$

when $X$ is an abelian surface, a K3 surface, a Fermat surface or a Delsarte surface (Gros and Suwa [8, Ch. IV,4]).

## 3. Abelian Surfaces

In this section, we shall confine ourselves to abelian surfaces defined over finite fields. By an abelian surface, we mean an abelian variety of dimension 2 . From 3.1 to 3.5 inclusive, k denotes a perfect field of characteristic $p > 0$ , and from 3.6 on $k = \mathbb{F}_q$ a finite field of $q = p^a$ elements. Let X be an abelian surface over k .

**3.1.** The canonical bundle $K_X$ is trivial. This implies that $NS(X_{\bar{k}})$ is torsion-free. For a prime $\ell \neq p$ , (2.2) combined with the torsion-free-ness of $NS(X_{\bar{k}})$ asserts that $H^{\cdot}(X_{\bar{k}}, \mathbb{Z}_\ell)$ is also torsion-free. Then by Künneth formula, $H^{\cdot}(X_{\bar{k}}, \mathbb{Z}_\ell) = \wedge H^1(X_{\bar{k}}, \mathbb{Z}_\ell)$ .

**3.2.** For $\ell = p$ , we have the corresponding facts for the crystalline cohomology groups of X .

The Hodge spectral sequence

$$E_1^{ij} = H^j(X, \Omega_X^i) \Longrightarrow H_{DR}^{i+j}(X/k)$$

degenerates at $E_1$ . Further, we have $\sum_{i+j=n} h^{ij}(X) = B_n(X)$ . From these two facts, we can deduce that $h_W^{ij}(X) = h^{ij}(X)$ for any pair $(i,j)$ with $i+j = n$ . Therefore

$$\dim_k H_{DR}^n(X/k) = \sum_{i+j=n} h^{ij}(X) = B_n(X) .$$

Hence by the universal coefficient theorem, $H^{\cdot}(X/W)$ is torsion-free. Then the Künneth formula asserts that $H^{\cdot}(X/W) = \wedge H^1(X/W)$ .

It is known (Mazur and Messing [22]) that $H^1(X/W)$ is isomorphic to the Dieudonné module of the p-divisible group $_{p^\infty}X$ associated to X .

The Hodge numbers $h^{ij}(X) = \dim H^j(X, \Omega_X^i)$ are computed in the following diagram:

| | $\mathcal{O}_X$ | $\Omega_X^1$ | $\Omega_X^2$ |
|---|---|---|---|
| $H^2$ | 1 | 2 | 1 |
| $H^1$ | 2 | 4 | 2 |
| $H^0$ | 1 | 2 | 1 |

**3.2.1. Definition.** The *p-rank* of $X$, denoted by $p\text{-rk}(X)$, is defined by

$$p\text{-rk}(X) := \dim_{\mathbb{F}_p} {}_pX(\bar{k}) .$$

**3.3.** The slope sequence of $H^1(X/W)$, and hence by (3.2), the slope sequence of $H^2(X/W)$, are completely determined by the $p$-rank of $X$. The results are tabulated as follows:

| p-rk(X) | slope seq. of $H^1(X/W)$ | slope seq. of $H^2(X/W)$ |
|---|---|---|
| 2 | 0, 0, 1, 1 | 0, 1, 1, 1, 1, 2 |
| 1 | 0, 1/2, 1/2, 1 | 1/2, 1/2, 1, 1, 3/2, 3/2 |
| 0 | 1/2, 1/2, 1/2, 1/2 | 1, 1, 1, 1, 1, 1 |

Observe that the $p\text{-rk}(X)$ is equal to the number of 0's in the slope sequence of $H^1(X/W)$.

**3.3.1. Definition.** An abelian surface $X$ is said to be *ordinary* (resp. *supersingular*) if the $p\text{-rk}(X)$ is equal to 2 (resp. 0).

**3.3.2. Remark.** If $X$ is an abelian variety of arbitrary dimension, we define $X$ to be *ordinary* if $p\text{-rk}(X) = \dim X$ and $X$ to be *supersingular* if $H^1(X/W)$ has the pure slope $1/2$. It follows that if $X$ is supersingular, then $p\text{-rk}(X) = 0$. However, the converse is not true if $\dim X \geq 3$.

**3.4.** Now we shall consider the formal Brauer group $\hat{Br}_{X/k}$ of $X$. Since $p_g(X) = 1$ and $Pic_{X/k}$ is smooth, $\hat{Br}_{X/k} = H^2(X, \hat{\mathbb{G}}_{m,X})$ is represented by a smooth formal group of dimension $p_g(X) = 1$ over $k$. Recall that $H^2(X, W\mathcal{O}_X)$ is the Cartier module of $\hat{Br}_{X/k}$ and that $H^2(X, W\mathcal{O}_X)_K = H^2(X/W)_K^{[0,1[}$. From these and (3.3), we see that the following conditions are all equivalent:

(i) $X$ is of Hodge-Witt type,

(ii) $\hat{Br}_{X/k}$ is p-divisible, and

(iii) $p\text{-rk}(X) \geq 1$.

When one of the above equivalent conditions is valid, h (= the height of $\widehat{Br}_{X/k}$) is finite, and

$$h = \begin{cases} 1 & \text{if } X \text{ is ordinary} \\ 2 & \text{if } X \text{ has } p\text{-rk}(X) = 1 . \end{cases}$$

On the other hand, if X is supersingular, $\widehat{Br}_{X/k}$ is unipotent.

**3.4.1. Remark.** Over the algebraic closure $\bar{k}$, the structure of $\widehat{Br}_{X/\bar{k}}$ is determined by its height:

$$\widehat{Br}_{X/\bar{k}} \simeq \begin{cases} \hat{\mathbb{G}}_{m,\bar{k}} & \text{if } h = 1 \\ G_{1,1} & \text{if } h = 2 \\ \hat{\mathbb{G}}_{a,\bar{k}} & \text{if } h = \infty . \end{cases}$$

**3.5.** The cohomology groups $H^j(X_{\bar{k}}, \mathbb{Z}_p(i))$ are computed in the following diagrams.

**3.5.1. Ordinary abelian surfaces.**

| | | |
|---|---|---|
| | | $H^4(X_{\bar{k}}, \mathbb{Z}_p(2)) = \mathbb{Z}_p$ |
| | $H^3(X_{\bar{k}}, \mathbb{Z}_p(1)) = \mathbb{Z}_p^2$ | $H^3(X_{\bar{k}}, \mathbb{Z}_p(2)) = \mathbb{Z}_p^2$ |
| $H^2(X_{\bar{k}}, \mathbb{Z}_p) = \mathbb{Z}_p$ | $H^2(X_{\bar{k}}, \mathbb{Z}_p(1)) = \mathbb{Z}_p^4$ | $H^2(X_{\bar{k}}, \mathbb{Z}_p(2)) = 0$ |
| $H^1(X_{\bar{k}}, \mathbb{Z}_p) = \mathbb{Z}_p^2$ | $H^1(X_{\bar{k}}, \mathbb{Z}_p(1)) = \mathbb{Z}_p^2$ | |
| $H^0(X_{\bar{k}}, \mathbb{Z}_p) = \mathbb{Z}_p$ | | |

**3.5.2. Abelian surfaces of $p\text{-rk}(X) = 1$ .**

| | | |
|---|---|---|
| | | $H^4(X_{\bar{k}}, \mathbb{Z}_p(2)) = \mathbb{Z}_p$ |
| | $H^3(X_{\bar{k}}, \mathbb{Z}_p(1)) = \mathbb{Z}_p$ | $H^3(X_{\bar{k}}, \mathbb{Z}_p(2)) = \mathbb{Z}_p$ |
| $H^2(X_{\bar{k}}, \mathbb{Z}_p) = 0$ | $H^2(X_{\bar{k}}, \mathbb{Z}_p(1)) = \mathbb{Z}_p^2$ | $H^2(X_{\bar{k}}, \mathbb{Z}_p(2)) = 0$ |
| $H^1(X_{\bar{k}}, \mathbb{Z}_p) = \mathbb{Z}_p$ | $H^1(X_{\bar{k}}, \mathbb{Z}_p(1)) = \mathbb{Z}_p$ | |
| $H^0(X_{\bar{k}}, \mathbb{Z}_p) = \mathbb{Z}_p$ | | |

**3.5.3. Supersingular abelian surfaces.**

| | | |
|---|---|---|
| | | $H^4(X_{\bar{k}}, \mathbb{Z}_p(2)) = \mathbb{Z}_p$ |
| | $H^3(X_{\bar{k}}, \mathbb{Z}_p(1)) = \bar{k}$ | $H^3(X_{\bar{k}}, \mathbb{Z}_p(2)) = 0$ |
| $H^2(X_{\bar{k}}, \mathbb{Z}_p) = 0$ | $H^2(X_{\bar{k}}, \mathbb{Z}_p(1)) = \mathbb{Z}_p^6$ | $H^2(X_{\bar{k}}, \mathbb{Z}_p(2)) = 0$ |
| $H^1(X_{\bar{k}}, \mathbb{Z}_p) = 0$ | $H^1(X_{\bar{k}}, \mathbb{Z}_p(1)) = 0$ | |
| $H^0(X_{\bar{k}}, \mathbb{Z}_p) = \mathbb{Z}_p$ | | |

**3.6.** We shall now turn our discussions to the computations of the special values of $P_2(X, q^{-s})$ as $s \to 0, 1$ .

First of all, the formula in (2.8) together with the facts in (3.1) , (3.2) and (3.5) yields

$$P_2(X, 1) = \pm \, |H^2(X, \mathbb{Q}/\mathbb{Z})| \, .$$

The Tate conjecture is valid for $X$ (Tate [42]) , so that the Picard number $\rho(X)$ is equal to the multiplicity $\rho_1$ of $q$ as a reciprocal root of $P_2(X, T) = 0$ .

Furthermore, the Artin-Tate formula for X takes the following simpler form:

$$\left[\frac{P_2(X,T)}{(1-qT)^{\rho(X)}}\right]_{T=1/q} = \frac{|Br(X)||\det NS(X)|}{q} .$$

**3.7.** Now we shall recall the Honda-Tate theory for abelian varieties of arbitrary dimension over finite fields (Honda [9]; Cf. Waterhouse [44]). The Honda-Tate theory will provide data for the calculation of special values of $P_2(X;T)$ at $T = 1/q$. Let $k = \mathbb{F}_q$ be a finite field with $q = p^a$ elements.

An algebraic integer $\pi$ is called a *Weil number* (for $k = \mathbb{F}_q$) if all conjugates of $\pi$ have the complex absolute value $q^{1/2}$. If $\pi$ is a Weil number, put $\beta = \pi + q/\pi$. Then $\beta$ is totally real and $|\beta| \le 2\sqrt{q}$ in every embedding. Conversely, any given $\beta$ with these properties gives rise to a Weil number $\pi$ as a root of $X^2 - \beta X + q = 0$.

The category of abelian varieties over k is semi-simple up to k-isogeny.

**3.7.1. Definition.** An abelian variety X over k is said to be k-*simple* if X defines a simple object in the category of abelian varieties over k up to k-isogeny, that is, if X contains no non-trivial abelian subvarieties over k.

It is known that X is k-simple if and only if the endomorphism algebra $\mathcal{E} := End_k(X) \otimes_{\mathbb{Z}} \mathbb{Q}$ is a division algebra over $\mathbb{Q}$. If X is k-simple, then $P_1(X;T) = Q(X;T)^e$ where $Q(X;T)$ is a $\mathbb{Q}$-irreducible polynomial and $e^2$ gives the rank of the division algebra $\mathcal{E}$ over its center. Let $\pi$ be a reciprocal root of $P_1(X;T) = 0$. Then $\pi$ is a Weil number and $\mathbb{Q}(\pi)$ is the center of $\mathcal{E}$ (cf. Waterhouse [44]).

The classification theorem of Honda and Tate on k-simple abelian varieties over $k = \mathbb{F}_q$ is formulated as follows:

**3.7.2. Theorem.** *The map*

$$X \longrightarrow \text{a reciprocal root of } P_1(X;T) = 0$$

*gives a bijection*

$$\left\{\begin{array}{l}\text{isogeny classes of } k\text{-simple}\\ \text{abelian varieties over}\\ k = \mathbb{F}_q\end{array}\right\} \xrightarrow{\sim} \left\{\begin{array}{l}\text{conjugacy classes}\\ \text{of Weil numbers}\\ \text{for } k = \mathbb{F}_q\end{array}\right\}$$

**3.8.** Here let $X$ be an abelian variety of dimension $1$, i.e., an elliptic curve defined over $k = \mathbb{F}_q$ with $q = p^a$, and we shall elaborate on the classification theorem of Honda and Tate for elliptic curves. (Waterhouse [44, Th. 4.1].)

**3.8.1. Proposition.** *Isogeny classes of ordinary elliptic curves $X$ over $k = \mathbb{F}_q$ are in 1-1 correspondence with the set of $\mathbb{Q}$-irreducible polynomials $P_1(X;T) = 1 - \beta T + qT^2 \in \mathbb{Z}[T]$ where $\beta = \pi + q/\pi$ with a Weil number $\pi$, $|\beta| < 2\sqrt{q}$ and $(\beta, p) = 1$.*

**3.8.2. Proposition.** *Isogeny classes of supersingular elliptic curves $X$ over $k = \mathbb{F}_q$ are in 1-1 correspondence with the set of polynomials $P_1(X;T) \in \mathbb{Z}[T]$ which are tabulated as follows. (A Weil number $\pi$ has the property that $(\pi/\sqrt{q})^n = 1$ for some $n \in \mathbb{N}$.)*

| $P_1(X;T)$ | Weil number(s) | n | $k = \mathbb{F}_q$ with $q = p^a$ |
|---|---|---|---|
| $(1-\sqrt{q}T)^2$ | $\sqrt{q}$ | 1 | $2 \mid a$ |
| $(1+\sqrt{q}T)^2$ | $-\sqrt{q}$ | 2 | $2 \mid a$ |
| $1+qT^2$ | $\pm\sqrt{q}$ | 4 | $2 \nmid a$ or $2 \mid a$, $p \not\equiv 1 \pmod 4$ |
| $1+\sqrt{q}T+qT^2$ | $\frac{-1\pm i\sqrt{3}}{2}\sqrt{q}$ | 3 | $2 \mid a$, $p \not\equiv 1 \pmod 3$ |
| $1-\sqrt{q}T+qT^2$ | $\frac{1\pm i\sqrt{3}}{2}\sqrt{q}$ | 6 | $2 \mid a$, $p \not\equiv 1 \pmod 3$ |
| $1+\sqrt{2q}T+qT^2$ | $\frac{-1\pm i}{2}\sqrt{q}$ | 8 | $2 \nmid a$, $p = 2$ |
| $1-\sqrt{2q}T+qT^2$ | $\frac{1\pm i}{2}\sqrt{q}$ | 8 | $2 \nmid a$, $p = 2$ |
| $1+\sqrt{3q}T+qT^2$ | $\frac{-\sqrt{3}\pm i}{2}\sqrt{q}$ | 12 | $2 \nmid a$, $p = 3$ |
| $1-\sqrt{3q}T+qT^2$ | $\frac{\sqrt{3}\pm i}{2}\sqrt{q}$ | 12 | $2 \nmid a$, $p = 3$ |

**3.9.** Now we pass onto the classification of abelian surfaces over $k = \mathbb{F}_q$ up to k-isogeny.

**3.9.1. Proposition.** *Isogeny classes of k-simple abelian surfaces* X *over* $k = \mathbb{F}_q$ *with* $p\text{-rk}(X) \geq 1$ *are in* 1-1 *correspondence with the set of* $\mathbb{Q}$*-irreducible polynomials*

$$P_1(X;T) = 1 + bT + cT^2 + qbT^3 + q^2T^4 \in \mathbb{Z}[T]$$

*with* $b = -\operatorname{Tr}(\beta)$ , $c = 2q + \operatorname{Nr}(\beta)$ *where* $\beta$ *is a real quadratic integer which is not rational and with absolute value* $< 2\sqrt{q}$ *in any real embedding. (Here* $\operatorname{Tr}(\beta)$ *(resp.* $\operatorname{Nr}(\beta)$*) denotes the trace (resp. norm) of* $\beta$ *.)*

*Furthermore, the following hold:*

X *is ordinary if and only if* $(\beta, p) = 1$ , i.e., $(\operatorname{Tr}(\beta), p) = (\operatorname{Nr}(\beta), p) = 1$ , *and*

X *is of* $p\text{-rk}(X) = 1$ *if and only if* $(\operatorname{Tr}(\beta), p) = 1$ *but* $p \mid \operatorname{Nr}(\beta)$ .

**3.9.2. Proposition.** *Isogeny classes of* k*-simple supersingular abelian surfaces* X *over* $k = \mathbb{F}_q$ *with* $q = p^a$ *are in* 1-1 *correspondence with the set of polynomials* $P_1(X;T)$ *which are tabulated as follows. (A Weil number* $\pi$ *has the property that* $(\pi/\sqrt{q})^n = 1$ *for some* $n \in \mathbb{N}$ *.)*

| $P_1(X;T)$ | Weil number(s) | n | $k = \mathbb{F}_q$ with $q = p^a$ |
|---|---|---|---|
| $(1-qT^2)^2$ | $\pm\sqrt{q}$ | 2 | $2 \nmid a$ |
| $(1+qT^2)^2$ | $\pm i\sqrt{q}$ | 4 | $2 \mid a$ , $p \equiv 1 \pmod 4$ |
| $(1+\sqrt{q}T+qT^2)^2$ | $\frac{-1\pm i\sqrt{3}}{2}\sqrt{q}$ | 3 | $2 \mid a$ , $p \equiv 1 \pmod 3$ |
| $(1+\sqrt{q}T+qT^2)^2$ | $\frac{1\pm i\sqrt{3}}{2}\sqrt{q}$ | 6 | $2 \mid a$ , $p \equiv 1 \pmod 3$ |
| $1+q^2T^4$ | $\frac{\pm1\pm i}{2}\sqrt{q}$ | 8 | $2 \nmid a$ , $p \neq 2$ or $2 \mid a$ , $p \not\equiv 1 \pmod 4$ |
| $1+qT^2+q^2T^4$ | $\frac{\pm\sqrt{3}\pm i}{2}\sqrt{q}$ | 12 | $2 \nmid a$ , $p \neq 3$ or $2 \mid a$ , $p \not\equiv 1 \pmod{12}$ |

(continued)

| $P_1(X;T)$ | Weil number(s) | n | $k = \mathbb{F}_q$ with $q = p^a$ |
|---|---|---|---|
| $1+\sqrt{q}T+qT^2$ $+q\sqrt{q}T^3+q^2T^4$ | $e^{\pm 2\pi i/5}\sqrt{q}$ $e^{\pm 4\pi i/5}\sqrt{q}$ | 5 | $2\|a$ , $p \not\equiv 1 \pmod 5$ |
| $1-\sqrt{q}T+qT^2$ $-q\sqrt{q}T^3+q^2T^4$ | $e^{\pm \pi i/5}\sqrt{q}$ $e^{\pm 3\pi i/5}\sqrt{q}$ | 10 | $2\|a$ , $p \not\equiv 1 \pmod 5$ |
| $1+\sqrt{2q}T+qT^2$ $+q\sqrt{2q}T^3+q^2T^4$ | $e^{\pm 5\pi i/12}\sqrt{q}$ $e^{\pm 7\pi i/12}\sqrt{q}$ | 24 | $2\nmid a$ , $p = 2$ |
| $1-\sqrt{2q}T+qT^2$ $-q\sqrt{2q}T^3+q^2T^4$ | $e^{\pm \pi i/12}\sqrt{q}$ $e^{\pm 11\pi i/12}\sqrt{q}$ | 24 | $2\nmid a$ , $p = 2$ |

**3.9.3. Remark.** Let $X$ be a k-simple abelian surface over $k = \mathbb{F}_q$. If $p\text{-rk}(X) \geq 1$, then $X_{\bar{k}}$ remains $\bar{k}$-simple. However, if $p\text{-rk}(X) = 0$, a k-simple supersingular abelian surface $X$ is not $\bar{k}$-simple, indeed, $X_{\bar{k}}$ is $\bar{k}$-isogenous to a product of two supersingular elliptic curves over $\bar{k}$. (Oort [31, Th.4.2].)

**3.10.** Let $X$ be an abelian surface over $k = \mathbb{F}_q$. The congruence zeta-function of $X$ has the form

$$Z(X;T) = \frac{P_1(X,T)\, P_1(X,qT)}{(1-T)\, P_2(X,T)\, (1-q^2T)}$$

where $P_i(X,T) \in 1 + T\mathbb{Z}[T]$ $(i = 1,2)$ with $\deg P_1 = 4$ and $\deg P_2 = 6$.

For any prime $\ell \neq p$, we have

$$H^2(X_{\bar{k}}, \mathbb{Q}_\ell) = \wedge^2 H^1(X_{\bar{k}}, \mathbb{Q}_\ell),$$

and therefore, $P_2(X;T)$ is completely determined from $P_1(X;T)$: Indeed if

$$P_1(X;T) = \prod_{i=1}^{4} (1 - \alpha_i T) \in \mathbb{C}[T]$$

then

$$P_2(X;T) = \prod_{i<j} (1 - \alpha_i\alpha_j T) \in \mathbb{C}[T].$$

The polynomial $P_1(X;T)$ is completely determined relative to $k = \mathbb{F}_q$ from the propositions in (3.8) and (3.9). Indeed, the following facts are at our disposal:

(1) If $X$ is $k$-isogenous to the product $E_1 \times E_2$ of elliptic curves $E_i$ $(i = 1,2)$, then

$$P_1(X;T) = P_1(E_1;T)\,P_1(E_2;T) .$$

(2) If $X$ is $k$-simple, then $P_1(X;T) = Q(X;T)^e$ with a $\mathbb{Q}$-irreducible polynomial $Q(X;T)$ and $e = 1$ or $2$. The cases $e = 2$ occur only when $X$ is supersingular.

These facts give rise to the assertions in (3.11), (3.12) and (3.13) below as corollaries.

**3.11.** Now we shall compute the special values of $P_2(X;T)$ at $T = 1$.

**3.11.1. Corollary.** *Let $X$ be an abelian surface over $k = \mathbb{F}_q$ which is $k$-isogenous to the product $E_1 \times E_2$ of elliptic curves $E_i$ over $k$ with Weil numbers $\pi_i$ and $\beta_i = \pi_i + q/\pi_i \in \mathbb{Z}$ $(i = 1,2)$. Then*

$$P_2(X;1) = (1-q)^2(1-2q^2+q^4-\beta_1\beta_2+q\beta_1^2+q\beta_2^2-q^2\beta_1\beta_2) .$$

*In particular, if $E_1$ and $E_2$ are $k$-isogenous, then*

$$P_2(X;1) = (1-q)^4(1+2q+q^2-\beta^2) \quad \textit{with } \beta := \beta_1 = \beta_2 ,$$

*and furthermore, if $E_1$ and $E_2$ are $k$-isogenous with the Weil number $\pi = \pm\sqrt{q} \in \mathbb{Z}$, then*

$$P_2(X;1) = (1-q)^6 .$$

**3.11.2. Corollary.** *Let $X$ be a $k$-simple abelian surface over $k = \mathbb{F}_q$ with a Weil number $\pi$.*

*(1) If $e = 1$, then*

$$P_2(X;1) = (1-q)^2\{1-2q^2+q^4-(1+2q+q^2)Nr(\beta)+qTr(\beta)^2\}$$

where $\beta = \pi + q/\pi$ is purely real quadratic.

*(2) If* $e = 2$ *, then*

$$P_2(X;1) = \begin{cases} (1-q)^2(1+q)^4 & \text{if } \pi = \pm\sqrt{q} \notin \mathbb{Z}, \\ (1-q)^4(1+2q+q^2-\beta^2) & \text{otherwise} \end{cases}$$

*where* $\beta = \pi + q/\pi$ *is rational.*

**3.12.** The Picard number $\rho(X)$ of an abelian surface $X$ over $k = \mathbb{F}_q$ is computed and the results are summarized as follows.

**3.12.1. Corollary.** *Let* $X$ *be an abelian surface over* $k = \mathbb{F}_q$ *with* $q = p^a$ *. Assume that* $X$ *is* $k$*-isogenous to the product* $E_1 \times E_2$ *of elliptic curves* $E_i$ $(i = 1,2)$ *over* $k$ *.*

*(1) If* $E_1$ *is not* $k$*-isogenous to* $E_2$ *, then* $\rho(X) = 2$ *.*

*(2) If* $E_1$ *is* $k$*-isogenous to* $E_2$ *, then* $\rho(X) = 4$ or $6$ *.*

*In particular,* $\rho(X) = 6$ *if and only if* $E_1$ *and* $E_2$ *have the Weil number* $\pi = \pm\sqrt{q} \in \mathbb{Z}$ *.*

**3.12.2. Corollary.** *Let* $X$ *be an abelian surface over* $k = \mathbb{F}_q$ *with* $q = p^a$ *. Assume that* $X$ *is* $k$*-simple. Then* $\rho(X) = 2$ *or* $4$ *.*

*In particular,* $\rho(X) = 4$ *if and only if* $e = 2$ *and* $X$ *has a Weil number* $\pi$ *different from* $\pm\sqrt{q} \notin \mathbb{Z}$ *.*

**3.12.3. Remark.** We should include some comments on the Picard number $\rho(X_{\bar{k}})$ of an abelian surface $X$ over $\bar{k}$ . $\rho(X_{\bar{k}})$ is subject to the Igusa-Artin-Mazur inequality $\rho(X_{\bar{k}}) \le \dim H^2(X/W)_K$[1] . The validity of the Tate conjecture implies that $\rho(X_{\bar{k}}) \equiv 0 \pmod 2$. Combining (2.7) , (3.3) , (3.12.1) and (3.12.2), we can derive the following facts:

$X_{\bar{k}}$ is supersingular if and only if $\rho(X_{\bar{k}}) = 6$ .

If $X_{\bar{k}}$ is of p-rk(X) = 1 , then $\rho(X_{\bar{k}}) = 2$ .

If $X_{\bar{k}}$ is ordinary, then $\rho(X_{\bar{k}}) = 2$ or $4$ . Conversely, $\rho(X_{\bar{k}}) = 4$ if and only if $X$ is $\bar{k}$-isogenous to the product $E \times E$ of an ordinary elliptic curve $E$ .

**3.13.** Now we shall evaluate the polynomial $P_2(X;T)$ at $T = 1/q$.

**3.13.1. Proposition (The Artin-Tate formula).** *Let* $X$ *be an abelian surface over* $k = \mathbb{F}_q$ *with* $q = p^a$. *Suppose that* $X$ *is* k-*isogenous to the product* $E_1 \times E_2$ *of elliptic curves* $E_i$ *over* $k$ *with Weil numbers* $\pi_i$ *and* $\beta_i = \pi_i + q/\pi_i$ $(i = 1,2)$. *Then*

$$\left[\frac{P_2(X;T)}{(1-qT)^{\rho(X)}}\right]_{T=1/q}$$

$$= \begin{cases} (\beta_1 - \beta_2)^2/q & \text{if } E_1 \text{ is not } k\text{-isogenous to } E_2, \\ 4 - \beta^2/q & \text{if } E_1 \text{ is } k\text{-isogenous to } E_2 \quad (\beta := \beta_1 = \beta_2), \\ 1 & \text{if } E_1 \text{ is } k\text{-isogenous to } E_2 \text{ and} \\ & \pi = \pm\sqrt{q} \in \mathbb{Z} \quad (\pi := \pi_1 = \pi_2). \end{cases}$$

**3.13.2. Proposition.** *Let* $X$ *be an abelian surface over* $k = \mathbb{F}_q$ *with* $q = p^a$. *Suppose that* $X$ *is* k-*simple with a Weil number* $\pi$ and $\beta = \pi + q/\pi$. *Then*

$$\left[\frac{P_2(X;T)}{(1-qT)^{\rho(X)}}\right]_{T=1/q}$$

$$= \begin{cases} [\mathrm{Tr}(\beta)^2 - 4\,\mathrm{Nr}(\beta)]/q & \text{if } \beta \notin \mathbb{Q} \\ 4 - \beta^2/q & \text{if } \beta \in \mathbb{Q}. \end{cases}$$

**3.14.** We should point out that $P_1(X;T)$ (and hence $P_2(X;T)$) are k-isogeny invariant. Hence by the Artin-Tate formula in (3.6), the product $|\mathrm{Br}(X)||\det \mathrm{NS}(X)|$ is an k-isogeny invariant, however, the individual pieces, $|\mathrm{Br}(X)|$, and $\det \mathrm{NS}(X)$, are definitely not.

Let $\mathrm{End}_k(X)$ be the endomorphism ring of $X$. Then $\mathrm{End}_k(X)$ is an order in the semi-simple algebra $\mathcal{E} = \mathrm{End}_k(X) \otimes_{\mathbb{Z}} \mathbb{Q}$ over $\mathbb{Q}$. Then the discriminant of $\mathrm{End}_k(X)$ determines $\det \mathrm{NS}(X)$. In particular, if $X$ is k-isomorphic to the product $E \times E$ of an elliptic curve $E$, then

$$\det \mathrm{NS}(X) = \mathrm{disc}\,\mathrm{End}_k(X).$$

For the following examplary cases, we shall compute $|Br(X)|$ and $\det NS(X)$ individually. Note that $\det NS(X) < 0$ by the Hodge index theorem.

We briefly recall the results of Waterhouse which will be relevant to the subsequent discussions.

**3.14.1. Theorem.** (*Waterhouse* [44 , Th. 4.2 *and* Th. 4.5)] . *Let* $\mathcal{E}$ *be the endomorphism algebra of an elliptic curve over* $k = \mathbb{F}_q$ *with a Weil number* $\pi$ . *Then the orders of* $\mathcal{E}$ *which occur as endomorphism rings of elliptic curves in the class are the following:*

*(1) If* E *is ordinary, all orders containing* $\pi$ ,

*(2) If* E *is supersingular with all endomorphisms defined over* k , *the maximal orders ; and if* E *is supersingular with not all endomorphisms defined over* k , *the orders which contain* $\pi$ *and with conductors prime to* $p = \mathrm{char}(k)$ .

*Furthermore, if* R *is an order in* $\mathbb{Q}(\pi)$ *which is realized as the endomorphism ring of an elliptic curve over* k , *then the number of* k-*isomorphism classes of elliptic curves with endomorphism rings* R *is equal to the class number of* R .

**3.14.2. Examples.** Let $X = E \times E$ where E is an elliptic curve over $k = \mathbb{F}_9$ $(q = 3^2)$ .

With the isogeny classes indexed by $\beta \in \mathbb{Z}$ (cf. (3.8)) the results on the computation of $|Br(X)|$ and $\det NS(X)$ are listed as follows:

(1) $\beta = 0$ , $4q-\beta^2 = 36$ , $\pi = \pm 3\sqrt{-1}$ . E is supersingular with not all endomorphisms defined over k . $\mathbb{Z}[\pi]$ cannot be realized as it has conductor 3 ; $\mathbb{Z}[\sqrt{-1}]$ is the maximal order in $\mathbb{Q}(\sqrt{-1})$ and there is a class of elliptic curves with $\mathbb{Z}[\sqrt{-1}]$ as the endomorphism ring. For this class

$$\det NS(X) = -4 \quad \text{and} \quad |Br(X)| = 3^2$$

(2) $\beta = \pm 1$ , $4q-\beta^2 = 35$ , $\pi = \frac{\mp 1 \pm \sqrt{-35}}{2}$ . E is ordinary. $\mathbb{Z}[\pi]$ is the maximal order in $\mathbb{Q}(\sqrt{-35})$ , and there are two classes of elliptic curves with $\mathbb{Z}[\pi]$ as the endomorphism rings. For both of these classes,

$$\det NS(X) = -35 \quad \text{and} \quad |Br(X)| = 1 .$$

(3) $\beta = \pm 2$ , $4q-\beta^2 = 32$ , $\pi = \mp 1 \pm 2\sqrt{-2}$ . E is ordinary. $\mathbb{Z}[\pi]$ is an order with conductor 2 in $\mathbb{Q}(\sqrt{-2})$ , and there is a class of elliptic curves with $\mathbb{Z}[\pi]$ as the endormorphism ring. For this class

$$\det NS(X) = -32 \quad \text{and} \quad |Br(X)| = 1 .$$

$\mathbb{Z}[\sqrt{-2}]$ is the maximal order in $\mathbb{Q}(\sqrt{-2})$ , and there is a class of elliptic curves with $\mathbb{Z}[\sqrt{-2}]$ as the endomorphism ring. For this class

$$\det NS(X) = -8 \quad \text{and} \quad |Br(X)| = 4 .$$

(4) $\beta = \pm 3$ , $4q-\beta^2 = 27$ , $\pi = \frac{\mp 3 \pm \sqrt{-3}}{2}$ . E is supersingular with not all endomorphisms defined over k . $\mathbb{Z}[\pi]$ cannot be realized as it has conductor 3 ; $\mathbb{Z}[\sqrt{-3}]$ is the maximal order in $\mathbb{Q}(\sqrt{-3})$ and there is a class of elliptic curves with $\mathbb{Z}[\sqrt{-3}]$ as the endomorphism ring. For this class,

$$\det NS(X) = -3 \quad \text{and} \quad |Br(X)| = 3^2 .$$

(5) $\beta = \pm 4$ , $4q-\beta^2 = 20$ , $\pi = \mp 2 \pm \sqrt{-5}$ . E is ordinary. $\mathbb{Z}[\pi] = \mathbb{Z}[\sqrt{-5}]$ is the maximal order in $\mathbb{Q}(\sqrt{-5})$ and there are two classes of elliptic curves with $\mathbb{Z}[\sqrt{-5}]$ as the endomorphism rings. For both classes,

$$\det NS(X) = -20 \quad \text{and} \quad |Br(X)| = 1 .$$

(6) $\beta = \pm 5$ , $4q-\beta^2 = 11$ , $\pi = \frac{\mp 5 \pm \sqrt{-11}}{2}$ . E is ordinary. $\mathbb{Z}[\pi] = \mathbb{Z}[\sqrt{-11}]$ is the maximal order in $\mathbb{Q}(\sqrt{-11})$ and there is a class of elliptic curves with $\mathbb{Z}[\pi]$ as the endomorphism ring. For this class,

$$\det NS(X) = -11 \quad \text{and} \quad |Br(X)| = 1$$

(7) $\beta = \pm 6$ , $4q-\beta^2 = 0$ , $\pi = \mp 3$ . E is supersingular with all endomorphisms defined over k . $\mathrm{End}_k(E)$ is the maximal order in the quaternion algebra, D , over $\mathbb{Q}$ ramifield only at 3 and $\infty$ . The class number of this order is 1 , and there is a class of elliptic curves having this order as the endomorphism ring. For this class,

$$\det NS(X) = -9 \quad \text{and} \quad |Br(X)| = 1 .$$

## 4. Kummer surfaces

In this section, we consider K3 surfaces. By a K3 surface $X$, we mean a projective smooth surface defined by the following conditions:

$$K_X \equiv 0 \quad \text{and} \quad H^1(X, \mathcal{O}_X) = 0 .$$

From 4.1 to 4.7 inclusive, the field $k$ of definition for $X$ is assumed to be a perfect field of characteristic $p > 0$, and from 4.8 to 4.12 inclusive we take $k$ to be a finite field $\mathbb{F}_q$ of $q = p^a$ elements.

**4.1.** The canonical bundle $K_X$ is trivial. This implies that $NS(X_{\bar{k}})$ is torsion-free. Let $\ell$ be a prime different from $p = \mathrm{char}(k)$. The condition $H^1(X, \mathcal{O}_X) = 0$ assures that $H^1(X_{\bar{k}}, \mathbb{Z}_\ell) = 0$. The fact that $NS(X_{\bar{k}})$ is torsion-free, together with (2.2) implies that $H^2(X_{\bar{k}}, \mathbb{Z}_\ell)$ is also torsion-free. Moreover, from these facts and the Poincaré duality, it follows that $H^3(X_{\bar{k}}, \mathbb{Z}_\ell) = 0$.

**4.2.** For a K3 surface $X$, $H^0(X, \Omega^1) = 0$ by Rudakov and Shafarevich [32] (see [16] for a short proof). The Hodge numbers $h^{ij}(X)$ are then determined and are tabulated as follow:

| | | | |
|---|---|---|---|
| $H^2$ | 1 | 0 | 1 |
| $H^1$ | 0 | 20 | 0 |
| $H^0$ | 1 | 0 | 1 |
| | $\mathcal{O}_X$ | $\Omega^1_X$ | $\Omega^2_X$ |

From this table, we can deduce that the Hodge spectral sequence

$$E_1^{ij} = H^j(X, \Omega^i) \Longrightarrow H^{i+j}_{DR}(X/k)$$

degenerates at $E_1$. Therefore, $\sum_{i+j=n} h^{ij}(X) = B_n(X)$.

Furthermore, the Hodge-Witt numbers $h_W^{ij}(X)$ coincide with the Hodge numbers $h^{ij}(X)$ for any pair $(i,j)$ with $i+j = n$.

From the condition $H^1(X,\mathcal{O}_X) = 0$, it follows that $H^1(X/W) = 0$. Then by the universal coefficient theorem, $H^2(X/W)$ is torsion-free. Finally, these two facts together with the Poincaré duality yield that $H^3(X/W) = 0$.

**4.3.** For a K3 surface $X$, the condition $H^1(X,\mathcal{O}_X) = 0$ implies that the Picard scheme $\mathrm{Pic}_{X/k}$ is smooth. Therefore the formal Brauer group $\hat{\mathrm{Br}}_{X/k} = H^2(X,\hat{\mathbb{G}}_{m,X})$ is represented by a smooth formal group over $k$ of $\dim \hat{\mathrm{Br}}_{X/k} = p_g(X) = 1$. The structure of $\hat{\mathrm{Br}}_{X/k}$ up to $\bar{k}$-isomorphism is determined by its height $h$.

If $h$ is finite, $\hat{\mathrm{Br}}_{X/k}$ is p-divisible and $H^2(X,W\mathcal{O}_X)$ is isomorphic to the Cariter module of $\hat{\mathrm{Br}}_{X/k}$. From this, it follows that $H^2(X,W\mathcal{O}_X)$ is free over $W$ of rank $h$, and that $H^2(X,W\mathcal{O}_X)$ is an F-crystal with slope $1-\frac{1}{h}$. The slope sequence of $H^2(X/W)$ is given as follows:

$$1-\frac{1}{h}\ ,\ 1\ ,\ 1+\frac{1}{h}$$

with multiplicities $h$, $22-2h$ and $h$, respectively. Therefore, $h$ must be subject to the inequality $1 \leq h \leq 10$.

If $h$ is infinite, $H^2(X/W)$ is an F-crystal with the pure slope 1, and $\hat{\mathrm{Br}}_{X/\bar{k}} \simeq \hat{\mathbb{G}}_{a,\bar{k}}$.

**4.3.1. Definition.** A K3 surface $X$ is said to be *ordinary* if $h = 1$. A K3 surface $X$ is said to be *supersingular* if $h = \infty$ (Artin [1]).

**4.4.** The cohomology groups $H^j(X_{\bar{k}},\mathbb{Z}_p(i))$ are computed in the following diagrams.

**4.4.1. Ordinary K3 surfaces.**

| | | |
|---|---|---|
| | | $H^4(X_{\bar{k}}, \mathbb{Z}_p(2)) = \mathbb{Z}_p$ |
| | $H^3(X_{\bar{k}}, \mathbb{Z}_p(1)) = 0$ | $H^3(X_{\bar{k}}, \mathbb{Z}_p(2)) = 0$ |
| $H^2(X_{\bar{k}}, \mathbb{Z}_p) = \mathbb{Z}_p$ | $H^2(X_{\bar{k}}, \mathbb{Z}_p(1)) = \mathbb{Z}_p^{20}$ | $H^2(X_{\bar{k}}, \mathbb{Z}_p(2)) = \mathbb{Z}_p$ |
| $H^1(X_{\bar{k}}, \mathbb{Z}_p) = 0$ | $H^1(X_{\bar{k}}, \mathbb{Z}_p(1)) = 0$ | |
| $H^0(X_{\bar{k}}, \mathbb{Z}_p) = \mathbb{Z}_p$ | | |

**4.4.2. K3 surfaces with height $h$ , $2 \le h < 10$ .**

| | | |
|---|---|---|
| | | $H^4(X_{\bar{k}}, \mathbb{Z}_p(2)) = \mathbb{Z}_p$ |
| | $H^3(X_{\bar{k}}, \mathbb{Z}_p(1)) = 0$ | $H^3(X_{\bar{k}}, \mathbb{Z}_p(2)) = 0$ |
| $H^2(X_{\bar{k}}, \mathbb{Z}_p) = 0$ | $H^2(X_{\bar{k}}, \mathbb{Z}_p(1)) = \mathbb{Z}_p^{22-2h}$ | $H^2(X_{\bar{k}}, \mathbb{Z}_p(2)) = 0$ |
| $H^1(X_{\bar{k}}, \mathbb{Z}_p) = 0$ | $H^1(X_{\bar{k}}, \mathbb{Z}_p(1)) = 0$ | |
| $H^0(X_{\bar{k}}, \mathbb{Z}_p) = \mathbb{Z}_p$ | | |

**4.4.3. Supersingular K3 surfaces.**

| | | |
|---|---|---|
| | | $H^4(X_{\bar{k}}, \mathbb{Z}_p(2)) = \mathbb{Z}_p$ |
| | $H^3(X_{\bar{k}}, \mathbb{Z}_p(1)) = \bar{k}$ | $H^3(X_{\bar{k}}, \mathbb{Z}_p(2)) = 0$ |
| $H^2(X_{\bar{k}}, \mathbb{Z}_p) = 0$ | $H^2(X_{\bar{k}}, \mathbb{Z}_p(1)) = \mathbb{Z}_p^{22}$ | $H^2(X_{\bar{k}}, \mathbb{Z}_p(2)) = 0$ |
| $H^1(X_{\bar{k}}, \mathbb{Z}_p) = 0$ | $H^1(X_{\bar{k}}, \mathbb{Z}_p(1)) = 0$ | |
| $H^0(X_{\bar{k}}, \mathbb{Z}_p) = \mathbb{Z}_p$ | | |

**4.5. Definition.** Let $A$ be an abelian surface defined over $k$ of characteristic $p \neq 2$. Let $\iota : A \to A$ be an involution of $A$. The *Kummer surface* $X$ associated to $A$, denoted by $Km(A)$, is obtained from the quotient $A/\iota$ by blowing up the 16 singular points corresponding to the points of order 2 on $A$. It is known that $X = Km(A)$ is a K3 surface.

Let ${}_2A = \{ a \in A \mid 2a = o_A \}$ denote the group of points of order 2 on $A$ where $o_A$ denotes the identity of $A$.. For each $a \in {}_2A$, let $E_a$ denote the smooth rational curve on X corresponding to $a$.

**4.6. Proposition.** *Let* $X = Km(A)$ *be a Kummer surface over* $k$ *of characteristic* $p \neq 2$. *Then the formal Brauer groups* $\widehat{Br}_{X/k}$ *and* $\widehat{Br}_{A/k}$ *are both represented by one-dimensional formal groups over* $k$, *and furthermore,*

$$\widehat{Br}_{X/k} \simeq \widehat{Br}_{A/k} \quad (k\text{-}isomorphic) .$$

**4.6.1. Corollary.** *Let* $X = Km(A)$ *defined over* $k$ *of characteristic* $p \neq 2$. *Then the following assertions hold.*

| $A$ | $X$ | $\widehat{Br}_{X/\bar{k}} \simeq \widehat{Br}_{A/\bar{k}}$ |
|---|---|---|
| ordinary | ordinary | $\hat{\mathbb{G}}_{m,\bar{k}}$ |
| $p\text{-}rk(X) = 1$ | height 2 | $G_{1,1}$ |
| supersingular | supersingular | $\hat{\mathbb{G}}_{a,\bar{k}}$ |

**4.7. Proposition.** *Let* $X = Km(A)$ *defined over* $k$ *of characteristic* $p \neq 2$. *Then*

$$\rho(X_{\bar{k}}) = 16 + \rho(A_{\bar{k}}) .$$

From here on, we take $k$ to be a finite field $\mathbb{F}_q$ with $q = p^a$ elements.

**4.8.** Let $X$ be a K3 surface defined over a finite field $k = \mathbb{F}_q$ where $q = p^a$. The congruence zeta-function of $X$ is of the form

$$Z(X;T) = \frac{1}{(1-T)\, P_2(X;T)\, (1-q^2T)}$$

where $P_2(X;T) \in \mathbb{Z}[T]$ with $\deg P_2 = 22$ , and

$$P_2(X;T) = \prod_{i=1}^{22} (1 - \alpha_i T) \in \mathbb{C}[T] \quad \text{with} \quad |\alpha_i| = q \ .$$

Therefore, the reciprocal roots of $P_2(X;T) = 0$ are of the form

$$\alpha_1 , \alpha_2 , \ldots , \alpha_{11} , q/\alpha_1 , q/\alpha_2 , \ldots , q/\alpha_{11} .$$

**4.9.** We shall now turn our discussions on to the evaluation of $P_2(X;T)$ at $T = 1$ .

The formula in (2.8) together with the facts stated in (4.1) and (4.3) give rise to a formula:

$$P_2(X;1) = \pm|H^2(X,\mathbb{Q}/\mathbb{Z})| \ .$$

Henceforth we shall confine ourselves to Kummer surfaces defined over finite fields $k = \mathbb{F}_q$ of characteristic different from 2 .

**4.10.** We shall determine the polynomial $P_2(X;T)$ for $X = \mathrm{Km}(A)$ . First of all, we know from the construction that

$$P_2(X;T) = R(X;T)\ P_2(A;T)$$

with some $R(X;T) \in \mathbb{Z}[T]$ of $\deg R = 16$ . $R(X;T)$ may be determined as follows. Let $\Gamma = \mathrm{Gal}(\bar{k}/k)$ be the Galois group of $\bar{k}$ over $k$ . Then $\Gamma$ acts on ${}_2A$ and hence it induces an action on $\sum_{a\in {}_2A} \mathbb{Z}\, E_a \subset NS(X_{\bar{k}})$ . This gives rise to a representation $\Gamma \longrightarrow GL(16,\mathbb{Z})$ . Then

$$R(X;T) = \det(1 - \phi^{-1}T \ ; \ GL(16,\mathbb{Z})) \ .$$

Furthermore, over the extension $k({}_2A)$ this representation becomes trivial. Therefore,

$$R(X;T) = \prod_{i=1}^{16} (1 - q\xi_i T) \in \mathbb{C}[T]$$

where $\xi_i$ $(i = 1,\ldots,16)$ are roots of unity.

The possibilities for $R(X;T)$ are listed in the table below. Here $F$ denotes the image of the topological generator $\phi$ of $\Gamma = \mathrm{Gal}(\bar{k}/k)$ under the representation $\Gamma \to GL(16,\mathbb{Z})$ . Possible choices for $R(X;T)$

depend on the polynomial $P_1(A;T) = 1 + bT + cT^2 + bqT^3 + q^2T^4$ (cf. (3.9)) and $R(X;T)$ are tabulated as follows:

| $P_1(A;T)$ | order of $F$ | $R(X;T)$ |
|---|---|---|
| $2\nmid b$ and $2\nmid c$ | 5 | $(1-qT)(1-q^5T^5)^3$ |
| $2\mid b$ and $2\nmid c$ | 3 | $(1-qT)(1-q^3T^3)^5$ |
| | 6 | $(1-qT)(1-q^3T^3)(1-q^6T^6)^2$ |
| $2\nmid b$ and $2\mid c$ | 3 | $(1-qT)^4(1-q^3T^3)^4$ |
| | 6 | $(1-qT)^2(1-q^2T^2)(1-q^3T^3)^2(1-q^6T^6)$ |
| $2\mid b$ and $2\mid c$ | 1 | $(1-qT)^{16}$ |
| | 2 | $(1-qT)^4(1-q^2T^2)^6$ |
| | 2 | $(1-qT)^8(1-q^2T^2)^4$ |
| | 4 | $(1-qT)^2(1-q^2T^2)(1-q^4T^4)^3$ |
| | 4 | $(1-qT)^4(1-q^2T^2)^2(1-q^4T^4)^2$ |

Note that $R(X;T)$ is not necessarily an isogeny invariant if the degree of the isogeny in question is even.

We shall explain how $R(X;T)$ is obtained by an example. The representation $\Gamma \longrightarrow GL(16,\mathbb{Z})$ gives rise to a representation $\Gamma \longrightarrow GL(4,\mathbb{F}_2)$. Let $\tilde{F}$ denote the image of $\phi$ under this representation. Then the characteristic polynomial of $\tilde{F}^{-1}$ with respect to ${}_2A(\bar{k}) \simeq \mathbb{F}_2^4$ coincides with $P_1(A;T) \pmod 2$. Therefore, for instance, if $2\nmid b$ and $2\nmid c$, then

$$\det(1-\tilde{F}^{-1}T;{}_2A) = P_1(A;T) \pmod 2 = 1+T+T^2+T^3+T^4 .$$

Hence $\tilde{F}$ is of order $5$, and hence so is $F$. Thus $|{}_2A(\mathbb{F}_{q^5})| = 16$ and $|{}_2A(\mathbb{F}_q)| = 1$. This gives rise to $R(X;T) = (1-qT)(1-q^5T^5)^3$.

**4.11.** let $X = Km(A)$ . Since $X$ has an elliptic fibration over $\mathbb{P}^1_k$ , the Tate conjecture is true by Artin and Swinnerton-Dyer [2] . Or the Tate conjecture can be proved more simply as follows: Since there is a dominant rational map $A \longrightarrow X$ and since the Tate conjecture is valid for $A$ by Tate [42] , the assertion (1) in (2.4) guarantees the validity of the Tate conjecture for $X$ .

The Artin-Tate formula for $X$ takes the following simple form

$$\left[\frac{P_2(X;T)}{(1-qT)^{\rho(X)}}\right]_{T=1/q} = \frac{|Br(X)||det\ NSX(X)}{q} .$$

**4.12. Example.** Let $A = E \times E$ be an abelian surface discussed in Example (3.14.2) . Let $X = Km(A)$ be the Kummer surface associated to $A$ . Then with the notations of (3.14.2) in force, we have the following results

(1) $(\beta = 0\ ,\ \mathbb{Z}[\sqrt{-1}])$ . Then

$P_2(X;T) = (1-qT)^4(1-q^2T^2)^6\ P_2(A;T)\ ,\ \rho(X) = 14$ ,

$|Br(X)||det\ NS(X)| = 2^6.36$, and furthermore,

$det\ NS(X) = -2^6.4$ and $|Br(X)| = 3^2$ .

(2) $(\beta = \pm 1\ ,\ \mathbb{Z}[\sqrt{-35}])$ . Then

$P_2(X;T) = (1-qT)(1-q^3T^3)^5\ P_2(A;T)\ ,\ \rho(X) = 10$ ,

$|Br(X)||det\ NS(X)| = 3^5.35$ and furthermore

$det\ NS(X) = -3^5.35$ and $|Br(X)| = 1$ .

(3) $(\beta = \pm 2\ ,\ \mathbb{Z}[\mp 1 \pm 2\sqrt{-2}])$ . Then

$P_2(X;T) = (1-qT)^4(1-q^2T^2)^6\ P_2(A;T)\ ,\ \rho(X) = 14$ ,

$|Br(X)||det\ NS(X)| = 2^6.32$ , and furthermore

$det\ NS(X) = -2^6.32$ and $|Br(X)| = 1$ .

$(\beta = \pm 2\ ,\ \mathbb{Z}[\sqrt{-2}])$ . Then

$P_2(X;T) = (1-qT)^{16}\ P_2(A;T)\ ,\ \rho(X) = 20$ ,

$|Br(X)||det\ NS(X) = 32$ , and furthermore,

$det\ NS(X) = 2^2(-8)$ , and $|Br(X)| = 1$ .

(4) $(\beta = \pm 3\ ,\ \mathbb{Z}[\sqrt{-3}])$ . Then

$$P_2(X;T) = (1-qT)(1-q^3T^3)^5\ P_2(A;T)\ ,\ \rho(X) = 10\ ,$$

$|Br(X)||\det NS(X)| = 3^5.27$ , and furthermore

$$\det NS(X) = -3^5.3 \quad \text{and} \quad |Br(X)| = 3^2\ .$$

(5) $(\beta = \pm 4\ ,\ \mathbb{Z}[\sqrt{-5}])$ . Then

$$P_2(X;T) = (1-qT)^4(1-q^2T^2)^6\ P_2(A;T)\ ,\ \rho(X) = 14\ ,$$

$|Br(X)||\det NS(X)| = 2^6.20$ , and furthermore

$$\det NS(X) = -2^6.20 \quad \text{and} \quad |Br(X)| = 1\ .$$

(6) $(\beta = \pm 5\ ,\ \mathbb{Z}[\sqrt{-11}])$ . Then

$$P_2(X;T) = (1-qT)(1-q^3T^3)^5\ P_2(A;T)\ ,\ \rho(X) = 10\ ,$$

$|Br(X)||\det NS(X)| = 3^5.11$ , and furthermore

$$\det NS(X) = -3^5.11 \quad \text{and} \quad |Br(X)| = 1\ .$$

(7) $(\beta = \pm 6$ , the maximal order in $D)$ . Then

$$P_2(X;T) = (1-qT)^{16}\ P_2(A;T)\ ,\ \rho(X) = 22\ ,$$

$|Br(X)||\det NS(X)| = 9$ , and furthermore

$$\det NS(X) = -9 \quad \text{and} \quad |Br(X)| = 1\ .$$

**4.13. Remark.** In this section, we assume that $k$ is algebraically closed field of characteristic $p \neq 2$ , and shall discuss relations between the Néron-Severi group of an abelian surface $A$ and that of the associated Kummer surface $X = Km(A)$ .

First we recall results of Shioda [34]. Shioda has obtained the identity

$$\det NS(X) = 2^{\nu} \det NS(A)$$

for some integer $\nu \geq 0$ . Further, he has shown that

(1) If $A$ is supersingular, then $\nu = 0$ ([loc. cit., Prop. 3.4])

and that

(2) $\nu = 22 - \rho(X) = 6 - \rho(A)$ provided that there is a lifting $A'$ of $A$ to characteristic zero such that $\rho(A') = \rho(A)$ ([loc. cit., Prop. 3.2]).

If $A$ is ordinary (resp. of $p\text{-rk}(X) \geq 1$) , then there is the canonical (resp. a quasi-canonical) lifting $A'$ to characteristic zero with $\rho(A') = \rho(A)$ . Therefore, we can remove the assumption in (2) .

## 5. Fermat surfaces

In this section, we shall consider the Fermat surface $X$ of degree $m$ defined over a finite field $k = \mathbb{F}_q$ of characteristic $p > 0$ with $(m,p) = 1$. $X$ is defined by the equation

$$X_0^m + X_1^m + X_2^m + X_3^m = 0$$

in the projective space $\mathbb{P}_k^3$. Throughout the section, we assume that $k = \mathbb{F}_q$ contains all the $m^{th}$ roots of unity. This is equivalent to the condition that $q \equiv 1 \pmod m$.

For the detailed accounts of this section, the reader should refer to Shioda [37] and Suwa and Yui [40].

5.1. For a smooth hypersurface $X$ in $\mathbb{P}_k^3$, $\mathrm{Pic}^0_{X/k} = 0$ and $NS(X_{\bar{k}})$ is torsion-free. Let $\ell$ be a prime different from $p = \mathrm{char}(k)$. Then these two facts imply that $H^1(X_{\bar{k}}, \mathbb{Z}_\ell) = H^3(X_{\bar{k}}, \mathbb{Z}_\ell) = 0$ and that $H^2(X_{\bar{k}}, \mathbb{Z}_\ell)$ is torsion-free. (Deligne [5].)

5.2. For a smooth hypersurface $X$ in $\mathbb{P}_k^3$, the Hodge spectral sequence

$$E_1^{ij} = H^j(X, \Omega_X^i) \Longrightarrow H_{DR}^{i+j}(X/k)$$

degenerates at $E_1$. Therefore, $\sum_{i+j=n} h^{ij}(X) = B_n(X)$. Furthermore, $H^1(X/W) = H^3(X/W) = 0$ and $H^2(X/W)$ is torsion-free. (Deligne [5].)

Therefore, the Hodge-Witt numbers $h_W^{ij}(X)$ coincide with the Hodge numbers $h^{ij}(X)$ for any pair $(i,j)$ with $i+j = n$, and they are tabulated as follows:

| | $\mathcal{O}_X$ | $\Omega_X^1$ | $\Omega_X^2$ |
|---|---|---|---|
| $H^2$ | $\frac{(m-1)(m-2)(m-3)}{6}$ | $0$ | $1$ |
| $H^1$ | $0$ | $\frac{m(2m^2-6m+7)}{3}$ | $0$ |
| $H^0$ | $1$ | $0$ | $\frac{(m-1)(m-2)(m-3)}{6}$ |

**5.3.** Here we shall fix the necessary notations.

$\mu_m = \{ \xi \in \mathbb{C}^\times \mid \xi^m = 1 \}$ = the group of $m^{th}$ roots of unity.

$G = G_m^2 = \mu_m^4/\text{diagonals} = \{ g = (\xi_0, \xi_1, \xi_2, \xi_3) \in \mu_m^4 \}/\text{diagonals} \subset \mathrm{Aut}(X)$

$\hat{G} = \hat{G}_m^2$ = the character group of $G$

$\simeq \{ a = (a_0, a_1, a_2, a_3) \mid a_i \in \mathbb{Z}/m\mathbb{Z} , \sum_{i=0}^{3} a_i \equiv 0 \pmod m \}$

under the correspondence $G \times \hat{G} \longrightarrow \mathbb{Q}(\xi_m) : (g,a) \longrightarrow a(g) = \prod_{i=0}^{3} \xi_i^{a_i}$

where $\xi_m$ denotes a primitive $m^{th}$ root of unity.

$\mathfrak{A} = \mathfrak{A}_m^2 = \{ a = (a_0, a_1, a_2, a_3) \in \hat{G} \mid a_i \not\equiv 0 \pmod m$ for every $i \}$ .

The group $(\mathbb{Z}/m\mathbb{Z})^\times$ acts on $\mathfrak{A}$ by componentwise multiplication.

$A = [a]$ = the $(\mathbb{Z}/m\mathbb{Z})^\times$-orbit of $a \in \mathfrak{A}$ .

$L = \mathbb{Q}(\xi_m)$ , and $L_A = \mathbb{Q}(\xi_m^d)$ where $d = \gcd(m,a)$ .

**5.4.** Now we recall the definition of Fermat motives, due to Shioda [37] , associated to the Fermat surface $X$ of degree $m$ .

For any $a \in \hat{G}$ , define

$$p_a = \frac{1}{|G|} \sum_{g \in G} a(g)^{-1} g = \frac{1}{m^4} \sum_{g \in G} a(g)^{-1} g$$

and for $A = [a]$ ,

$$p_A = \sum_{a \in A} p_a = \frac{1}{m^4} \sum_{g \in G} \mathrm{Tr}_{L_A/\mathbb{Q}}(a(g)^{-1}) g .$$

Then $p_a$ and $p_A$ are element of the group ring $L[G]$ or $\mathbb{Z}[\frac{1}{m}][G]$ . Furthermore, $p_a$ and $p_A$ are idempotents (projectors), that is

$$p_a \cdot p_b = \begin{cases} p_a & \text{if } a=b \\ 0 & \text{if } a \neq b \end{cases} , \quad \sum_{a \in \hat{G}} p_a = 1 ,$$

and

$$p_A \cdot p_B = \begin{cases} p_A & \text{if } A=B \\ 0 & \text{if } A \neq B \end{cases} , \quad \sum_{A \in O(\hat{G})} p_A = 1$$

where $O(\hat{G})$ denotes the set of $(\mathbb{Z}/m)^\times$-orbits in $\hat{G}$ .

Therefore, the pair $M_A = (X, p_A)$ defines a motive over $k$ with coefficients in $\mathbb{Z}[\frac{1}{m}]$ (cf. Manin [20]), called the *Fermat submotive* of $X$ corresponding to the $(\mathbb{Z}/m)^{\times}$-orbit $A$ in $\hat{G}$ (cf. Shioda [37], p. 125). If there is no danger of ambiguity, we simply call $M_A = (X, p_A)$ a *Fermat 2-motive* of $X$.

We call the decomposition $X = \bigoplus_{A \in O(\hat{G})} M_A$ the *motivic decomposition* of $X$.

**5.4.1 Remark.** The field of definition of $M_A$ is the prime field $\mathbb{F}_p$ (or $\mathbb{Q}$) (cf. Suwa and Yui [40]).

**5.5.** Let $R$ be a commutative ring with the identity element $1$, in which $m$ is invertible, and let $\mathcal{F}$ be a contravariant functor from a category of varieties over $k$ to the category of R-modules. For a Fermat 2-motive $M_A = (X, p_A)$, define

$$\mathcal{F}(M_A) = \mathrm{Im}[(p_A)_* : \mathcal{F}(X) \longrightarrow \mathcal{F}(X)] .$$

Examples of such a functor $\mathcal{F}$ are given as follows.

**5.5.1.** For any prime $\ell \neq p$, the $\ell$-adic étale cohomology groups $H^{\cdot}(X, \mathbb{Q}_\ell(i))$, $i \in \mathbb{Z}$; moreover, if $\ell$ is relatively prime to $m$, $H^{\cdot}(X. \mathbb{Z}/\ell^r(i))$, $H^{\cdot}(X, \mathbb{Z}_\ell(i))$, $H^{\cdot}(X, \mathbb{Q}_\ell/\mathbb{Z}_\ell(i))$, $i \in \mathbb{Z}$.

**5.5.2.** The de Rham cohomology groups $H^{\cdot}_{DR}(X/k)$, or the Hodge spectral sequence $E_1^{ij} = H^j(X, \Omega^i_X) \Rightarrow H^{i+j}_{DR}(X/k)$.

**5.5.3.** The crystalline cohomology groups $H^{\cdot}(X/W_n)$, $H^{\cdot}(X/W)$, $H^{\cdot}(X/W)_K$, or the slope spectral sequences $E_1^{ij} = H^j(X, W_n\Omega^i_X) \Rightarrow H^{\cdot}(X/W_n)$, and $E_1^{ij} = H^j(X, W\Omega^i_X) \Rightarrow H^{\cdot}(X/W)$ (cf. Illusie [13, Ch. II]).

**5.5.4.** The logarithmic Hodge-Witt cohomology groups $H^{\cdot}(X, \mathbb{Z}/p^r(i)) = H^{\cdot - i}(X, W_r\Omega^i_{X,\log})$, $H^{\cdot}(X, \mathbb{Z}_p(i)) = \varprojlim_R H^{\cdot}(X, \mathbb{Z}/p^r(i))$ and

$H^{\cdot}(X, \mathbb{Q}_p/\mathbb{Z}_p(i)) = \varinjlim_{\underline{p}} H^{\cdot}(X, \mathbb{Z}/p^r(i))$ , $i \in \mathbb{N}$ (cf. Illusie [12 , Ch. IV.3], or Gros and Suwa [8, Ch. I]) .

**5.6.** Let $M_A$ be a Fermat 2-motive over $k$ . Then $H^2(M_A/W)$ defines an F-crystal. By the *slopes and the Newton polygon* of $M_A$ , we mean those of the F-crystal $(H^2(M_A/W), F)$ , respectively.

Now we make the following definitions.

**5.7. Definition.** Let $M_A$ be a Fermat 2-motive over $k$ .

(1) $M_A$ is said to be *ordinary* if the Newton polygon and the Hodge polygon of $M_A$ coincide.

(2) $M_A$ is said to be *supersingular* if the Newton polygon has the pure slope 1 .

**5.8. Definition.** Let $M_A$ be a Fermat 2-motive over $k$ . $M_A$ is said to be *of Hodge-Witt type* if $H^2(M_A, WO_X)$ is a W-module of finite type.

Noting that the motivic decomposition of $X$ commutes with the cohomology functors $H^2(\ /W)$ , $H^2(\ , WO_X)$ , we obtain the following assertion.

**5.9. Proposition.** *Let* $X$ *be a Fermat surface of degree* $m$ *over* $k$ .

(1) $X$ *is ordinary if and only if each Fermat 2-motive of* $X$ *is ordinary.*

(2) $X$ *is of Hodge-Witt type if and only if each Fermat 2-motive of* $X$ *is of Hodge-Witt type.*

(3) $X$ *is supersingular if and only if each Fermat 2-motive of* $X$ *is supersingular.*

Here we need some discussions on Jacobi sums.

**5.10.** Choose a character $\chi : k^{\times} \longrightarrow \mathbb{C}^{\times}$ of exact order $m$ and fix it once and for all. For each $\mathbf{a} = (a_0, a_1, a_2, a_3) \in \mathfrak{A}$ , let $j(\mathbf{a}) = j(\mathbf{a})_{q,\chi}$ denote the Jacobi sum (relative to $q$ and $\chi$ ) defined by

$$j(a) = \sum \chi(v_1)^{a_1} \chi(v_2)^{a_2} \chi(v_3)^{a_3}$$

where the sum is taken over all vectors $(v_1, v_2, v_3) \in (k^\times)^3$ subject to the linear relation $v_1+v_2+v_3 = -1$.

Weil [45 , 46] has shown that

$$P_2(X,T) = (1-qT) \prod_{a\in\mathfrak{U}} (1-j(a)T) .$$

Some properties of Jacobi sums are listed in what follow.

(1) $j(a)$ is an algebraic integer in the $m^{th}$ cyclotomic field $L = \mathbb{Q}(\xi_m)$ with the complex absolute value $q$.

(2) If $\sigma_t \in Gal(L/\mathbb{Q}) \simeq (\mathbb{Z}/m\mathbb{Z})^\times$ such that $\sigma_t(\xi_m) = \xi_m^t$ with $t \in (\mathbb{Z}/m\mathbb{Z})^\times$, then $j(a)^{\sigma_t} = j(ta)$ with $ta \in \mathfrak{U}$.

(3) Let $\mathfrak{p}$ be a prime ideal in $L$ over $p$ with $Norm_{L/\mathbb{Q}}(\mathfrak{p}) = p^f$ where $f$ denotes the order of $p$ mod $m$. Let $H = \left\{ p^i \text{ mod } m \mid 0 \le i < f \right\}$ and let $G$ mod $H = \{s_1, \ldots, s_t\}$. Then the ideal $(j(a))$ has the prime ideal decomposition of the form

$$(j(a)) = \mathfrak{p}^{\omega(a)}$$

with

$$\omega(a) = \sum_{i=1}^{t} A_H(s_i a)\sigma_t \in \mathbb{Z}[Gal(L/\mathbb{Q})] .$$

Here

$$A_H(s_i a) = \sum_{t\in H} \| s_i ta \| ,$$

and

$$\| s_i ta \| = \sum_{i=0}^{3} \langle \frac{s_i ta_i}{m} \rangle - 1$$

$\langle x \rangle$ being the fractional part of $x \in \mathbb{Q}/\mathbb{Z}$.

(4) Jacobi sums $j(a)$ are the reciprocal roots of $P_2(X,T) = 0$ in $\mathbb{C}$

**5.10.1. Remark.** The slopes of $M_A$ are given by $\{ A_H(a)/f \}_{a\in A}$ .

It is rather useful to have combinatorial characterizations of ordinary, of Hodge-Witt type and supersingular Fermat 2-motives, respectively, in fact, they are given in the following proposition.

**5.11. Proposition.** *Let* $M_A$ *be a Fermat* 2-*motive over* $k$ .

(1) $M_A$ *is ordinary if and only if* $\|pa\| = 0$ *for any* $a \in A$ *with* $\|a\| = 0$ .

(2) $M_A$ *is of Hodge-Witt type if and only if* $\|pa\| \leq 1$ *for any* $a \in A$ *with* $\|a\| = 0$ .

(3) $M_A$ *is supersingular if and only if* $A_H(a) = f$ *for any* $a \in A$ *with* $\|a\| = 0$ .

As a consequence of Proposition (5.11) , one obtains

**5.11.1. Corollary.** The following conditions are all equivalent:

(i) $M_A$ is ordinary and supersingular,

(ii) $M_A$ is of Hodge-Witt type and supersingular,

(iii) $\|a\| = n/2$ for every $a \in A$ .

The motivic decomposition $X = \oplus_A M_A$ corresponds to the factorization of $P_2(X;T)$ :

$$P_2(X;T) = (1-qT) \prod_{A \in O(\hat{G})} P_A(T)$$

where

$$P_A(T) := P(M_A;T) = \prod_{a \in A} (1 - j(a)T) \in \mathbb{Z}[T] .$$

The Tate conjecture is valid for X (Tate [41] and Shioda and Katsura [38]) . Therefore the Picard number $\rho(X) = \rho_1$ ,and the actual calculation of $\rho_1$ can be carried out by passing to Fermat 2-motives.

**5.12. Theorem.** *Let* $M_A$ *be a Fermat 2-motive over* k . *Then the following conditions are all equivalent.*

(i) $M_A$ *is supersingular.*

(ii) *There is a prime* $\ell \neq p$ *such that*

$$NS(M_{A,\bar{k}}) \otimes_{\mathbb{Z}} \mathbb{Q}_\ell \longrightarrow H^2(M_{A,\bar{k}}, \mathbb{Q}_\ell(1)) \quad \textit{is bijective.}$$

(iii) *For all primes* $\ell \neq p$ , $NS(M_{A,\bar{k}}) \otimes_{\mathbb{Z}} \mathbb{Q}_\ell \longrightarrow H^2(M_{A,\bar{k}}, \mathbb{Q}_\ell(1))$ *is bijective.*

(iv) $NS(M_{A,\bar{k}}) \otimes_{\mathbb{Z}} \mathbb{Q} \neq 0$ .

(v) $j(\mathbf{a})/q$ *is a root of unity for any* $\mathbf{a} \in A$ .

(vi) $j(\mathbf{a})/q$ *is a root of unity for some* $\mathbf{a} \in A$ .

**5.12.1. Corollary.** *We have*

$$\rho(X_{\bar{k}}) = 1 + \sum B_2(M_A) ,$$

*where the summation is taken over all the supersingular Fermat 2-motives* $M_A$ *of* X .

**5.12.2. Examples.** The Picard numbers $\rho(X_{\bar{k}})$ for the Fermat surface X of degree m , $4 \leq m \leq 25$ are computed as follows. (Cf. Shioda [35] .)

Here $k = \mathbb{F}_q$ of characteristic $p > 0$ , f the order of p mod m . ord (resp. ss ; resp. H-W) stands for ordinary (resp. supersingular ; resp. Hodge-Witt type).

| m | p mod m | f | $\rho(X_{\overline{k}})$ | |
|---|---|---|---|---|
| 4 | 1 | 1 | 20 | ord |
| | 3 | 2 | 22 | ss |
| 5 | 1 | 1 | 37 | ord |
| | 2,3 | 4 | 53 | |
| | 4 | 2 | 53 | ss |
| 6 | 1 | 1 | 86 | ord |
| | 5 | 2 | 106 | ss |
| 7 | 1 | 1 | 91 | ord |
| | 3,5 | 6 | 187 | ss |
| | 2,4 | 3 | 91 | H-W |
| | 6 | 2 | 187 | ss |
| 8 | 1 | 1 | 176 | ord |
| | 3 | 2 | 178 | |
| | 5 | 2 | 188 | |
| | 7 | 2 | 302 | ss |
| 9 | 1 | 1 | 217 | ord |
| | 2,5 | 6 | 457 | ss |
| | 4,7 | 3 | 313 | |
| | 8 | 2 | 457 | ss |
| 10 | 1 | 1 | 362 | ord |
| | 3,7 | 4 | 658 | ss |
| | 9 | 2 | 658 | ss |
| 11 | 1 | 1 | 271 | ord |
| | 2,6 | 10 | 911 | ss |
| | 7,8 | 10 | 911 | ss |
| | 3,4 | 5 | 391 | |
| | 5,9 | 5 | 391 | |
| | 10 | 2 | 911 | ss |

| m | p mod m | f | $\rho(X_{\overline{k}})$ | |
|---|---|---|---|---|
| 12 | 1 | 1 | 644 | ord |
| | 5 | 2 | 688 | |
| | 7 | 2 | 682 | |
| | 11 | 2 | 1222 | ss |
| 13 | 1 | 1 | 397 | ord |
| | 2,7 | 12 | 1597 | ss |
| | 6,11 | 12 | 1597 | ss |
| | 4,10 | 6 | 1597 | ss |
| | 5,8 | 4 | 1597 | ss |
| | 3,9 | 3 | 397 | |
| | 12 | 2 | 1597 | ss |
| 14 | 1 | 1 | 806 | ord |
| | 3,5 | 6 | 2042 | ss |
| | 9,11 | 3 | 1190 | |
| | 13 | 2 | 2042 | ss |
| 15 | 1 | 1 | 835 | ord |
| | 2,8 | 4 | 899 | |
| | 7,13 | 4 | 1427 | |
| | 4 | 2 | 899 | |
| | 11 | 2 | 931 | |
| | 14 | 2 | 2563 | ss |
| 16 | 1 | 1 | 872 | ord |
| | 3,11 | 4 | 994 | |
| | 5,13 | 4 | 1676 | |
| | 7 | 2 | 1094 | |
| | 9 | 2 | 994 | |
| | 15 | 2 | 3166 | ss |
| 17 | 1 | 1 | 721 | ord |
| | 3,6 | 16 | 3857 | ss |
| | 5,7 | 16 | 3857 | ss |
| | 10,12 | 16 | 3857 | ss |

| m | p mod m | f | $\rho(X_{\bar{k}})$ | |
|---|---|---|---|---|
| 17 | 11,14 | 16 | 3857 | ss |
| | 2,9 | 8 | 3857 | ss |
| | 8,15 | 8 | 3857 | ss |
| | 4,13 | 4 | 3857 | ss |
| | 16 | 2 | 3857 | ss |
| 18 | 1 | 1 | 1658 | ord |
| | 5,11 | 6 | 4642 | ss |
| | 7,13 | 3 | 2282 | |
| | 17 | 2 | 4642 | ss |
| 19 | 1 | 1 | 919 | ord |
| | 2,10 | 18 | 5527 | ss |
| | 3,13 | 18 | 5527 | ss |
| | 14,15 | 18 | 5527 | ss |
| | 4,5 | 9 | 2215 | |
| | 6,16 | 9 | 2215 | |
| | 9,17 | 9 | 2215 | |
| | 8,12 | 6 | 5527 | ss |
| | 7,11 | 3 | 919 | |
| | 18 | 2 | 5527 | ss |
| 20 | 1 | 1 | 1988 | ord |
| | 3,7 | 4 | 2766 | |
| | 13,17 | 4 | 4404 | |
| | 9 | 2 | 2668 | |
| | 11 | 2 | 2158 | |
| | 19 | 2 | 6518 | ss |
| 21 | 1 | 1 | 1573 | ord |
| | 2,11 | 6 | 2365 | |
| | 5,17 | 6 | 7621 | ss |
| | 10,19 | 6 | 5245 | |
| | 4,16 | 3 | 2293 | |
| | 8 | 2 | 1645 | |
| | 13 | 2 | 1885 | |
| | 20 | 2 | 7621 | ss |

| m | p mod m | f | $\rho(X_{\bar{k}})$ | |
|---|---|---|---|---|
| 22 | 1 | 1 | 1742 | ord |
| | 7,19 | 10 | 8842 | ss |
| | 13,17 | 10 | 8842 | ss |
| | 3,15 | 5 | 2822 | |
| | 5,9 | 5 | 2822 | |
| | 21 | 2 | 8842 | ss |
| 23 | 1 | 1 | 1387 | ord |
| | 5,14 | 22 | 10187 | ss |
| | 7,10 | 22 | 10187 | ss |
| | 11,21 | 22 | 10187 | ss |
| | 15,20 | 22 | 10187 | ss |
| | 17,19 | 22 | 10187 | ss |
| | 2,12 | 11 | 4027 | |
| | 3,8 | 11 | 4027 | |
| | 4,6 | 11 | 4027 | |
| | 9,18 | 11 | 4027 | |
| | 13,16 | 11 | 4027 | |
| | 22 | 2 | 10187 | ss |
| 24 | 1 | 1 | 3080 | ord |
| | 5 | 2 | 3328 | |
| | 7 | 2 | 3530 | |
| | 11 | 2 | 4330 | |
| | 13 | 2 | 3164 | |
| | 17 | 2 | 3964 | |
| | 19 | 2 | 4726 | |
| | 23 | 2 | 11662 | ss |
| 25 | 1 | 1 | 1657 | ord |
| | 2,13 | 20 | 13273 | ss |
| | 3,17 | 20 | 13273 | ss |
| | 8,22 | 20 | 13273 | ss |
| | 12,23 | 20 | 13273 | ss |
| | 4,19 | 10 | 13273 | ss |
| | 9,14 | 10 | 13273 | ss |
| | 6,21 | 5 | 4537 | |
| | 11,16 | 5 | 4537 | |
| | 7,18 | 4 | 13273 | ss |
| | 24 | 2 | 13273 | ss |

**5.13.** With the validity of the Tate conjecture, the Artin-Tate formula for $X$ is read as follows:

$$\prod_{j(a)\neq q}(1-j(a)/q) = \pm\frac{|Br(X)||\det NS(X)|}{q^{p_g(X)}} .$$

Further, the decomposition $X = \bigoplus_A M_A$ defines a factorization of the Artin-Tate formula for $X$. For each prime number $\ell$ with $(\ell, mp) = 1$,

$$\text{(a)}\quad |\prod_A(\prod_{\substack{a\in A\\ j(a)\neq q}}(1-\frac{j(a)}{q}))|_\ell^{-1} = \prod_A |Br(M_A)_{\ell\text{-tors}}||\det NS(M_A)\otimes_{\mathbb{Z}}\mathbb{Z}_\ell|$$

and

$$\text{(b)}\quad |\prod_A(\prod_{\substack{a\in A\\ j(a)\neq q}}(1-\frac{j(a)}{q}))|_p^{-1} = \prod_A\frac{|Br(M_A)_{p\text{-tors}}||\det NS(M_A)\otimes_{\mathbb{Z}}\mathbb{Z}_p|}{q^{p_g(M_A)}}$$

More precisely, we have for any prime $\ell$,

$$|\prod_{\substack{a\in A\\ j(a)\neq q}}(1-\frac{j(a)}{q})|_\ell^{-1} =$$

$$\text{(AT)}\quad \begin{cases} |Br(M_A)_{\ell\text{-tors}}||\det NS(M_A)\otimes_{\mathbb{Z}}\mathbb{Z}_\ell| & (\ell, mp)=1 \\ \dfrac{|Br(M_A)_{p\text{-tors}}||\det NS(M_A)\otimes_{\mathbb{Z}}\mathbb{Z}_p|}{q^{p_g(M_A)}} & \ell = p . \end{cases}$$

**5.13.1. Theorem.** (The Artin-Tate formula I) *Suppose that* $M_A$ *is supersingular. Then the following assertions hold.*

(1) $Br(M_A)_{\ell\text{-tors}} = \{0\}$ *and* $|\det NS(M_A)\otimes_{\mathbb{Z}}\mathbb{Z}_\ell| = 1$ *for each prime* $\ell$ *with* $(\ell, mp) = 1$ .

(2) $|Br(M_A)_{p\text{-tors}}||\det NS(M_A)\otimes_{\mathbb{Z}}\mathbb{Z}_p| = q^{p_g(M_A)}$ .

**Proof.** Note that for each prime $\ell$ with $(\ell, m) = 1$, we have

$$|\prod_{\substack{a\in A\\ j(a)\neq q}} (1 - \frac{j(a)}{q}))|_{\ell}^{-1} = 1 .$$

Then the results follow from (AT).

**5.13.2. Theorem.** (The Artin-Tate formula II) *Suppose that* $M_A$ *is not supersingular. Then the following assertions hold.*

(1) $|Br(M_A)_{\ell\text{-tors}}| = |P_A(1/q)|_{\ell}^{-1} = |\prod_{a\in A} (1 - \frac{j(a)}{q})|_{\ell}^{-1}$ *for each prime* $\ell$ *with* $(\ell, mp) = 1$.

(2) $|Br(M_A)_{p\text{-tors}}|/q^{p_g(M_A)} = |P_A(1/q)|_p^{-1} = |\prod_{a\in A} (1 - \frac{j(a)}{q})|_p^{-1}$.

**Proof.** Since $M_A$ is not supersingular, $j(a) \neq q$ for any $a \in A$ and $NS(M_A)\otimes_{\mathbb{Z}} \mathbb{Z}_{\ell} = 0$ for any prime $\ell$ with $(\ell, m) = 1$. Thus the assertion follows from (AT).

On the explicit determination of $|Br(X)|$ and $\det NS(X)$, we have the following results.

**5.14. Theorem.** *If* $X$ *is of Hodge-Witt type, then* $\det NS(X)$ *divides a power of* $m$.

**5.14.1. Corollary.** *If* $X$ *is of Hodge-Witt type, then* $\det NS(X_{\bar{k}})$ *divides a power of* $m$.

**5.15. Theorem.** *Let* $m$ *be a prime* $> 3$. *Let* $M_A$ *be a Fermat 2-motive of Hodge-Witt type over* $k = \mathbb{F}_q$. *Assume that* $M_A$ *is not supersingular. Then*

$$Nr_{L/\mathbb{Q}}(1-j(a)/q) = \prod_{a\in A} (1-j(a)/q) = \pm Bm^3/q^{w^0(M_A)},$$

*where* $B$ *is a positive integer which is a square possibly multiplied by a divisor of* $2m$, *and* $w^0(M_A) = p_g(M_A) - T^{02} = p_g(M_A)$.

**Proof.** Let $\ell$ be a prime such that $(\ell, m) = 1$. The assertion on the m-part is a consequence of a theorem of Iwasawa on the congruence(s) of Jacobi sums obtained by using the theory of cyclotomic fields (Shioda [37, Prop.

3.1]). The assertion on B follows from the fact that $|Br(M_A)_{\ell\text{-tors}}|$ is a square or twice a square (Milne [24]). Then Theorem (5.13.2) gives rise to the required formula.

**5.15.1. Examples.** The values of $Nr_{L/Q}(1-j(a)/q)$ for m prime ≤ 19 and $q = p$ or $p^2$ with $p < 500$ and $p \equiv 1 \pmod m$ are computed by D. Zagier and the results are tabulated as follows.

**m = 5**

**a** = (1 1 1 2) , $w^0(M_A) = 1$

| p | 11 | 31 | 41 | 61 | 71 | 101 | 131 | 151 | 181 | 191 | 211 |
|---|---|---|---|---|---|---|---|---|---|---|---|
| (q=p) B | 1 | 1 | 1 | 1 | 1 | 1 | 1 | $4^2$ | 1 | 1 | $5^2$ |
| ($q=p^2$) B | 1 | $11^2$ | $9^2$ | 1 | $19^2$ | $29^2$ | $11^2$ | $16^2$ | $11^2$ | $41^2$ | $5^2$ |
| p | 241 | 251 | 271 | 281 | 311 | 331 | 401 | 421 | 431 | 461 | 491 |
| (q=p) B | $4^2$ | $4^2$ | 1 | $5^2$ | 1 | 1 | 1 | $5^2$ | $4^2$ | $5^2$ | 1 |
| ($q=p^2$) B | $64^2$ | $16^2$ | $31^2$ | $55^2$ | $49^2$ | $61^2$ | $29^2$ | $95^2$ | $144^2$ | $5^2$ | $9^2$ |

**m = 7**

**a** = (1 1 1 4) , $w^0(M_A) = 2$

| p | 29 | 43 | 71 | 113 | 127 | 197 | 211 | 239 |
|---|---|---|---|---|---|---|---|---|
| (q=p) B | 1 | 1 | 1 | $13^2$ | $41^2$ | $13^2$ | $71^2$ | $83^2$ |
| ($q=p^2$) B | $97^2$ | $29^2$ | $41^2$ | $13^2 167^2$ | $41^2 43^2$ | $13^2 71^2$ | $13^2 71^2$ | $83^4$ |
| p | 281 | 337 | 379 | 421 | 449 | 463 | 491 | |
| (q=p) B | 1 | $29^2$ | $7^2$ | $127^2$ | $41^2$ | $71^2$ | $13^2$ | |
| ($q=p^2$) B | $883^2$ | $29^2 449^2$ | $7^2 2407^2$ | $83^2 127^2$ | $41^2 349^2$ | $71^2 967^2$ | $13^2 2437^2$ | |

$a = (1\ 1\ 2\ 3)$ , $w^0(M_A) = 1$

| p | 29 | 43 | 71 | 113 | 127 | 197 | 211 | 239 |
|---|---|---|---|---|---|---|---|---|
| (q=p) B | 1 | 1 | 1 | 1 | 1 | 1 | 1 | 1 |
| ($q=p^2$) B | 1 | $13^2$ | 1 | $29^2$ | $13^2$ | $13^2$ | $43^2$ | $41^2$ |
| p | 281 | 337 | 379 | 421 | 449 | 463 | 491 | |
| (q=p) B | 1 | 1 | $7^2$ | 1 | 1 | 1 | 1 | |
| ($q=p^2$) B | $83^2$ | $43^2$ | $7^2$ | $29^2$ | $43^2$ | $41^2$ | $113^2$ | |

**m = 11**

$a = (1\ 1\ 1\ 8)$ , $w^0(M_A) = 3$

| p | 23 | 67 | 89 | 199 | 331 | 353 |
|---|---|---|---|---|---|---|
| (q=p) B | 1 | $23^2$ | $67^2$ | $736^2$ | $89^2$ | $1849^2$ |
| ($q=p^2$) B | $43^2$ | $23^2 109^2$ | $67^2 2507^2$ | $736^4$ | $89^2 8579^2$ | $1849^2 4489^2$ |
| p | 397 | 419 | 463 | | | |
| (q=p) B | $736^2$ | $32^2$ | $67^2$ | | | |
| ($q=p^2$) B | $736^2 2144^2$ | $32^2 106976^2$ | $67^2 4423^2$ | | | |

$a = (1\ 1\ 2\ 7)$ , $w^0(M_A) = 2$

| p | 23 | 67 | 89 | 199 | 331 | 353 | 397 |
|---|---|---|---|---|---|---|---|
| (q=p) B | 1 | 1 | $23^2$ | $43^2$ | $32^2$ | $23^2$ | $109^2$ |
| ($q=p^2$) B | $197^2$ | $43^2$ | $23^2$ | $43^2$ | $32^4$ | $23^2 199^2$ | $109^2 307^2$ |
| p | 419 | 463 | | | | | |
| (q=p) B | $23^2$ | $23^2$ | | | | | |
| ($q=p^2$) B | $23^2 947^2$ | $23^2 1277^2$ | | | | | |

$a = (1\ 1\ 3\ 6)\ ,\ w^0(M_A) = 3$

| p | 23 | 67 | 89 | 199 | 331 | 353 |
|---|---|---|---|---|---|---|
| (q=p) B | $43^2$ | 1 | 1 | $67^2$ | $529^2$ | 1 |
| ($q=p^2$) B | $43^2$ | $1429^2$ | $7613^2$ | $67^2 19823^2$ | $529^2 22133^2$ | $3013^2$ |
| p | 397 | 419 | 463 | | | |
| (q=p) B | $461^2$ | $463^2$ | $989^2$ | | | |
| ($q=p^2$) B | $43^2 461^2$ | $463^2 38477^2$ | $989^2 36851^2$ | | | |

$a = (1\ 1\ 4\ 5)\ ,\ w^0(M_A) = 2$

| p | 23 | 67 | 89 | 199 | 331 | 353 | 397 |
|---|---|---|---|---|---|---|---|
| (q=p) B | 1 | 1 | 1 | $23^2$ | $23^2$ | 1 | $67^2$ |
| ($q=p^2$) B | $67^2$ | $89^2$ | $23^2$ | $23^2 967^2$ | $23^2$ | $43^2$ | $67^2 199^2$ |
| p | 419 | 463 | | | | | |
| (q=p) B | $67^2$ | 1 | | | | | |
| ($q=p^2$) B | $43^2 67^2$ | $857^2$ | | | | | |

$a = (1\ 2\ 3\ 5)\ ,\ w^0(M_A) = 1$

| p | 23 | 67 | 89 | 199 | 331 | 353 | 397 | 419 | 463 |
|---|---|---|---|---|---|---|---|---|---|
| (q=p) B | 1 | 1 | 1 | 1 | 1 | 1 | 1 | 1 | 1 |
| ($q=p^2$) B | 1 | $23^2$ | 1 | $43^2$ | $109^2$ | $23^2$ | $67^2$ | 1 | $89^2$ |

**m = 13**

a = (1 1 1 10) $w^0(M_A) = 4$

| p | 53 | 79 | 131 | 157 | 313 |
|---|---|---|---|---|---|
| (q=p) B | $103^2$ | $131^2$ | $103^2$ | $5^4 79^2$ | $131^2$ |
| ($q=p^2$) B | $103^2 131^2$ | $131^2 337^2$ | $103^2 72463^2$ | $5^4 79^2 74801^2$ | $131^2 296297^2$ |
| p | 443 | | | | |
| (q=p) B | $103^2$ | | | | |
| ($q=p^2$) B | $5^4 103^2 2729^2$ | | | | |

a = (1 1 2 9) , $w^0(M_A) = 3$

| p | 53 | 79 | 131 | 157 | 313 | 443 |
|---|---|---|---|---|---|---|
| (q=p) B | $53^2$ | 1 | $103^2$ | $3^6$ | $157^2$ | $5^4 23^2$ |
| ($q=p^2$) B | $53^2 547^2$ | $9697^2$ | $103^2 1117^2$ | $3^6 6917^2$ | $157^2 12403^2$ | $5^4 23^2 20047^2$ |

a = (1 1 3 8) , $w^0(M_A) = 3$

| p | 53 | 79 | 131 | 157 | 313 | 443 |
|---|---|---|---|---|---|---|
| (q=p) B | $53^2$ | $131^2$ | 1 | $3^6$ | $53^2$ | $911^2$ |
| ($q=p^2$) B | $53^2$ | $131^2 157^2$ | $17861^2$ | $3^6 1039^2$ | $53^2 35437^2$ | $911^2 25247^2$ |

a = (1 1 4 7) , $w^0(M_A) = 3$

| p | 53 | 79 | 131 | 157 | 313 | 443 |
|---|---|---|---|---|---|---|
| (q=p) B | $79^2$ | $5^4$ | $313^2$ | 1 | $1117^2$ | $2731^2$ |
| ($q=p^2$) B | $79^2 233^2$ | $5^4 4003^2$ | $313^2 2133^2$ | $5^4 1223^2$ | $1117^2 12611^2$ | $2731^2 24751^2$ |

$a = (1\ 1\ 5\ 6)$ , $w^0(M_A) = 2$

| p | 53 | 79 | 131 | 157 | 313 | 443 |
|---|---|---|---|---|---|---|
| (q=p) B | 1 | 1 | 1 | 1 | $103^2$ | 1 |
| ($q=p^2$) B | $131^2$ | $5^4$ | $1171^2$ | $571^2$ | $103^4$ | $3821^2$ |

$a = (1\ 2\ 3\ 7)$ , $w^0(M_A) = 2$

| p | 53 | 79 | 131 | 157 | 313 | 443 |
|---|---|---|---|---|---|---|
| (q=p) B | 1 | 1 | $53^2$ | $3^6$ | $5^4$ | 1 |
| ($q=p^2$) B | $937^2$ | 1 | $53^2$ | $3^6$ | $5^4$ | $53^2$ |

$a = (1\ 3\ 4\ 5)$ , $w^0(M_A) = 1$

| p | 53 | 79 | 131 | 157 | 313 | 443 |
|---|---|---|---|---|---|---|
| (q=p) B | 1 | 1 | 1 | 1 | 1 | 1 |
| ($q=p^2$) B | 1 | 1 | $5^4$ | 1 | $53^2$ | $337^2$ |

**m = 17**

$a = (1\ 1\ 1\ 14)$ , $w^0(M_A) = 5$

| p | 103 | 137 | 239 | 307 |
|---|---|---|---|---|
| (q=p) B | $101^2$ | $1361^2$ | $1871^2$ | $827833^2$ |
| ($q=p^2$) B | $101^2 353701^2$ | $1361^2 10403^2$ | $1871^2 927079^2$ | $276691^2 827833^2$ |
| p | 409 | 443 | | |
| (q=p) B | $239257^2$ | $2^8 31621^2$ | | |
| ($q=p^2$) B | $239257^2 13006157^2$ | $2^{16} 31621^2 244121^2$ | | |

$a = (1\ 1\ 2\ 13)$, $w^0(M_A) = 4$

| p | 103 | 137 | 239 | 307 |
|---|---|---|---|---|
| (q=p) B | $4523^2$ | $647^2$ | $2857^2$ | $41309^2$ |
| $(q=p^2)$ B | $1733^2 4523^2$ | $647^2 2617^2$ | $2857^2 252893^2$ | $373^2 41309^2$ |
| p | 409 | 443 | | |
| (q=p) B | $137^2$ | $4217^2$ | | |
| $(q=p^2)$ B | $137^2 752827^2$ | $4217^2 5237^2$ | | |

$a = (1\ 1\ 3\ 12)$, $w^0(M_A) = 4$

| p | 103 | 137 | 239 | 307 | 409 |
|---|---|---|---|---|---|
| (q=p) B | 1 | $307^2$ | $2^8 373^2$ | $4489^2$ | $103^2$ |
| $(q=p^2)$ B | $99041^2$ | $307^2 7207^2$ | $2^{16} 373^2 30703^2$ | $4489^2 71671^2$ | $103^2 6349397^2$ |
| p | 443 | | | | |
| (q=p) B | 1 | | | | |
| $(q=p^2)$ B | $20131297^2$ | | | | |

$a = (1\ 1\ 4\ 11)$, $w^0(M_A) = 3$

| p | 103 | 137 | 239 | 307 | 409 | 443 |
|---|---|---|---|---|---|---|
| (q=p) B | $137^2$ | $101^2$ | 1 | $577^2$ | $443^2$ | 1 |
| $(q=p^2)$ B | $101^2 137^2$ | $101^2 137^2$ | $307291^2$ | $577^2 919^2$ | $443^2 509^2$ | $15809^2$ |

$a = (1\ 1\ 5\ 10)$, $w^0(M_A) = 4$

| p | 103 | 137 | 239 | 307 |
|---|---|---|---|---|
| (q=p) B | $1733^2$ | $5099^2$ | $103^2$ | $443^2$ |
| $(q=p^2)$ B | $1733^2 24991^2$ | $5099^2 24991^2$ | $103^2 20569^2$ | $443^2 718013^2$ |

(continued)

| p | 409 | 443 |
|---|---|---|
| (q=p) B | $11287^2$ | $1021^2$ |
| (q=$p^2$) B | $271^2 11287^2$ | $1021^2 1339973^2$ |

$a = (1\ 1\ 6\ 9)$ , $w^0(M_A) = 5$

| p | 103 | 137 | 239 |
|---|---|---|---|
| (q=p) B | $1699^2$ | $2^8 1019^2$ | $161569^2$ |
| (q=$p^2$) B | $1699^2 1516127^2$ | $2^{16} 1019^2 52427^2$ | $70039^2 161569^2$ |
| p | 307 | 409 | 443 |
| (q=p) B | $1291^2$ | $40391^2$ | $38419^2$ |
| (q=$p^2$) B | $1291^2 4460903^2$ | $40391^2 15153323^2$ | $38419^2 1271771^2$ |

$a = (1\ 1\ 7\ 8)$ , $w^0(M_A) = 3$

| p | 103 | 137 | 239 | 307 | 409 | 443 |
|---|---|---|---|---|---|---|
| (q=p) | 1 | $443^2$ | $103^2$ | $2^8$ | $2143^2$ | 1 |
| (q=$p^2$) B | $10709^2$ | $443^2$ | $103^2 33457^2$ | $2^{16} 20161^2$ | $2143^2 10403^2$ | $350879^2$ |

$a = (1\ 2\ 3\ 11)$ , $w^0(M_A) = 2$

| p | 103 | 137 | 239 | 307 | 409 | 443 |
|---|---|---|---|---|---|---|
| (q=p) B | 1 | 1 | 1 | 1 | $509^2$ | $409^2$ |
| (q=$p^2$) B | $1733^2$ | 1 | $5711^2$ | $103^2$ | $509^2$ | $137^2 409^2$ |

$a = (1\ 2\ 4\ 10)$ , $w^0(M_A) = 3$

| p | 103 | 137 | 239 | 307 | 409 | 443 |
|---|---|---|---|---|---|---|
| (q=p) B | $67^2$ | $169^2$ | $103^2$ | 1 | $169^2$ | $577^2$ |
| (q=$p^2$) B | $67^2 9929^2$ | $169^2 3671^2$ | $103^2 1973^2$ | $5441^2$ | $169^2 19211^2$ | $577^2 2789^2$ |

$a = (1\ 2\ 6\ 8)$ , $w^0(M_A) = 3$

| p | 103 | 137 | 239 | 307 | 409 | 443 |
|---|---|---|---|---|---|---|
| (q=p) B | $2^8$ | 1 | 1 | $409^2$ | $307^2$ | $307^2$ |
| ($q=p^2$) B | $2^{16}$ | $120223^2$ | $647^2$ | $409^2 1429^2$ | $307^2 2617^2$ | $307^2 3673^2$ |

$a = (1\ 3\ 5\ 8)$ , $w^0(M_A) = 2$

| p | 103 | 137 | 239 | 307 | 409 | 443 |
|---|---|---|---|---|---|---|
| (q=p) B | 1 | 1 | 1 | 1 | $67^2$ | $2^8$ |
| ($q=p^2$) B | $307^2$ | $647^2$ | $101^2$ | $3061^2$ | $67^2 1123^2$ | $2^{16} 67^2$ |

$a = (2\ 3\ 5\ 7)$ , $w^0(M_A) = 1$

| p | 103 | 137 | 239 | 307 | 409 | 443 |
|---|---|---|---|---|---|---|
| (q=p) B | 1 | 1 | 1 | 1 | 1 | 1 |
| ($q=p^2$) B | 1 | $67^2$ | $137^2$ | 1 | $101^2$ | 1 |

**m = 19**

$a = (1\ 1\ 1\ 16)$ , $w^0(M_A) = 6$

| p | 191 | 229 | 419 |
|---|---|---|---|
| (q=p) B | $146869^2$ | $2983229^2$ | $5419903^2$ |
| ($q=p^2$) B | $146869^2 24775733^2$ | $2983229^2 32811859^2$ | $5419903^2 72819059^2$ |
| p | 457 | | |
| (q=p) B | $4247449^2$ | | |
| ($q=p^2$) B | $4247449^2 760845689^2$ | | |

$a = (1\ 1\ 2\ 15)$ , $w^0(M_A) = 4$

| p | 191 | 229 | 419 |
|---|---|---|---|
| (q=p) B | $4597^2$ | $2659^2$ | $37^2$ |
| (q=p²) B | $4597^2 12769^2$ | $2659^2 23939^2$ | $37^2 6773843^2$ |
| p | 457 | | |
| (q=p) B | $219983^2$ | | |
| (q=p²) B | $1217^2 219983^2$ | | |

$a = (1\ 1\ 3\ 14)$ , $w^0(M_A) = 4$

| p | 191 | 229 | 419 |
|---|---|---|---|
| (q=p) B | $4447^2$ | $151^2$ | $12577^2$ |
| (q=p²) B | $4447^2 183767^2$ | $151^2 518851^2$ | $191^2 12577^2$ |
| p | 457 | | |
| (q=p) B | $113^2$ | | |
| (q=p²) B | $113^2 145009^2$ | | |

$a = (1\ 1\ 4\ 13)$ , $w^0(M_A) = 5$

| p | 191 | 229 | 419 |
|---|---|---|---|
| (q=p) B | $23977^2$ | $224047^2$ | $20483^2$ |
| (q=p²) B | $23977^2 824029^2$ | $55633^2 224047^2$ | $20483^2 38759^2$ |
| p | 457 | | |
| (q=p) B | $346597^2$ | | |
| (q=p²) B | $346597^2 10244609^2$ | | |

$a = (1\ 1\ 5\ 12)$ , $w^0(M_A) = 5$

| p | 191 | 229 | 419 |
|---|---|---|---|
| (q=p) B | $17519^2$ | $683^2$ | $683^2$ |
| ($q=p^2$) B | $17519^2 10848659^2$ | $683^2 12769^2$ | $683^2 40038433^2$ |
| p | 457 | | |
| (q=p) B | $6991^2$ | | |
| ($q=p^2$) B | $6991^2 19966987^2$ | | |

$a = (1\ 1\ 6\ 11)$ , $w^0(M_A) = 4$

| p | 191 | 229 | 419 |
|---|---|---|---|
| (q=p) B | $113^2$ | $1787^2$ | $1787^2$ |
| ($q=p^2$) B | $113^2 59051^2$ | $1787^2 9539^2$ | $1787^2 3031981^2$ |
| p | 457 | | |
| (q=p) B | $3041^2$ | | |
| ($q=p^2$) B | $3041^2 36671^2$ | | |

$a = (1\ 1\ 7\ 10)$ , $w^0(M_A) = 5$

| p | 191 | 229 | 419 |
|---|---|---|---|
| (q=p) B | $1369^2$ | $110543^2$ | $253613^2$ |
| ($q=p^2$) B | $1369^2 9453869^2$ | $110543^2 293057^2$ | $253613^2 332653^2$ |
| p | 459 | | |
| (q=p) B | $97697^2$ | | |
| ($q=p^2$) B | $97697^2 3141877^2$ | | |

$a = (1\ 1\ 8\ 9)$ , $w^0(M_A) = 3$

| p | 191 | 229 | 419 | 457 |
|---|---|---|---|---|
| (q=p) B | 1 | $37^2$ | $571^2$ | 1 |
| ($q=p^2$) B | $6803^2$ | $37^2 64639^2$ | $571^2 5587^2$ | $154471^2$ |

$a = (1\ 2\ 3\ 13)$ , $w^0(M_A) = 3$

| p | 191 | 229 | 419 | 457 |
|---|---|---|---|---|
| (q=p) B | $1369^2$ | 1 | 1 | $227^2$ |
| ($q=p^2$) B | $151^2 1369^2$ | $227^2$ | $130759^2$ | $227^2 457^2$ |

$a = (1\ 2\ 4\ 12)$ , $w^0(M_A) = 4$

| p | 191 | 229 | 419 |
|---|---|---|---|
| (q=p) B | $5587^2$ | $343^2$ | $23939^2$ |
| ($q=p^2$) B | $5587^2 19037^2$ | $343^2 1685413^2$ | $4181^2 23939^2$ |
| p | 457 | | |
| (q=p) B | $11287^2$ | | |
| ($q=p^2$) B | $11287^2 266341^2$ | | |

$a = (1\ 2\ 5\ 11)$ , $w^0(M_A) = 3$

| p | 191 | 229 | 419 | 457 |
|---|---|---|---|---|
| (q=p) B | $37^2$ | $607^2$ | $1^2$ | $37^2$ |
| ($q=p^2$) B | $37^2 6689^2$ | $607^2 14327^2$ | $293057^2$ | $37^2 8093^2$ |

$a = (1\ 2\ 7\ 9)\ ,\ w^0(M_A) = 3$

| p | 191 | 229 | 419 | 457 |
|---|---|---|---|---|
| (q=p) B | $151^2$ | 1 | $113^2$ | $229^2$ |
| ($q=p^2$) B | $151^4$ | $23369^2$ | $113^2 1369^2$ | $229^2 23939^2$ |

$a = (1\ 3\ 7\ 8)\ ,\ w^0(M_A) = 2$

| p | 191 | 229 | 419 | 457 |
|---|---|---|---|---|
| (q=p) B | 1 | $37^2$ | 1 | $797^2$ |
| ($q=p^2$) B | $1597^2$ | $37^4$ | $2053^2$ | $37^2 797^2$ |

$a = (1\ 4\ 5\ 9)\ ,\ w^0(M_A) = 2$

| p | 191 | 229 | 419 | 457 |
|---|---|---|---|---|
| (q=p) B | 1 | $191^2$ | 1 | 1 |
| ($q=p^2$) B | $4181^2$ | $37^2 191^2$ | $797^2$ | $37^2$ |

$a = (3\ 4\ 5\ 7)\ ,\ w^0(M_A) = 4$

| p | 191 | 229 | 419 | 457 |
|---|---|---|---|---|
| (q=p) B | 1 | 1 | 1 | 1 |
| ($q=p^2$) B | 1 | $37^2$ | $569^2$ | $37^2$ |

## REFERENCES

[1] Artin, M.,
Supersingular K3 surfaces,
Ann. scient. Éc. Norm. Sup. $4^e$ série, t. 7 (1974), pp. 543-568.

[2] Artin, M., and Swinnerton-Dyer, P.,
The Tate-Šafarevič conjecture for pencils of elliptic curves on K3 surfaces,
Invent. Math. 20 (1973), pp. 279-296.

[3] Artin, M., and Mazur, B.,
Formal groups arising from algebraic varieties,
Ann. scient. Éc. Norm. Sup. $4^e$ série, t. 10 (1977), pp. 87-132

[4] Crew, R.,
On torsion in the slope spectral sequence,
Compositio Math. 56 (1985), pp. 79-86.

[5] Deligne, P.,
Cohomologie des intersections complètes,
SGA 7, exp. IX, Lecture Notes in Math. 340,
Springer-Verlag (1973), pp. 401-438.

[6] Ekedahl, T.,
Diagonal complexes and F-guage structure,
Prépublication, Université de Paris-Sud (1985).

[7] Faltings, G.,
Endlichkeitssätze für abelsche Varietäten über Zahlkörpern,
Invent. Math. 73 (1983), pp. 349-366.

[8] Gros, M., and Suwa, N.,
Application d'Abel-Jacobi p-adique et cycles algébriques,
Duke J. Math. (1988) (to appear).

[9] Honda, T.,
Isogeny classes of abelian varieties over finite fields,
J. Math. Soc. Japan 20 (1968), pp. 83-95.

[10] Illusie, L.,
Report on Crystalline Cohomology,
Proc. Symp. Pure Math. Vol. XXIX (1975), pp. 459-478.

[11] Illusie, L.,
Complexe de de Rham-Witt,
Journée de Géometrie algébrique de Rennes, Asterisque 63,
Soc. Math. France (1979), pp. 83-112.

[12] Illusie, L.,
Finiteness, duality and Künneth theorems in the cohomology of the Rham-Witt complex,
Proc. of the Japan-France conference in Algebraic Geometry, Tokyo-Kyoto, Lecture Notes in Math. 1016, Springer-Verlag (1983), pp. 20-72.

[13] Illusie, L.,
Complexe de Re Rham-Witt et cohomologie cristalline,
Ann. scient. Éc. Norm. Sup. 4e série, t. 12 (1979), pp. 501-661.

[14] Illusie, L., and Raynaud, M.,
Les suites spectrales associées au complexe de de Rham-Witt,
Publ. Math. IHES 57 (1983), pp. 73-212.

[15] Katz, N.,
Slope filtrations of F-crystals,
Journeé de Géometrie algébrique de Rennes, Asterisque 63, Soc. Math. France (1979), pp. 113-164.

[16] Lang, W., and Nygaard, N.,
A short proof of the Rydakov-Šafarevič Theorem,
Math. Ann. 251 (1980), pp. 171-173.

[17] Lichtenbaum, S.,
Zeta-functions of varieties over finite fields at s = 1,
Arithmetic and Geometry I, Prog. in Math. Vol. 35, Birkhäuser (1983), pp. 173-194.

[18] Lichtenbaum, S.,
Values of zeta functions at non-zero integers,
Number Theory, Lecture Notes in Math. 1068, Springer-Verlag (1984), pp. 127-138.

[19] Manin, Y.,
The theory of commutative formal groups over fields of finite characteristic,
Russian Math. Surveys 10 (1963), pp. 1-83.

[20] Manin, Y.,
Correspondences, motives and monoidal transformations,
Math, USSR-Sb. 77 (1970), pp. 475-506.

[21] Mazur, B.,
Frobenius and the Hodge filtration,
Bull. Amer. Math. Soc. 78 (1972) pp. 653-667.

[22] Mazur, B., and Messing. W.,
Universal extensions and one dimensional crystalline cohomology,
Lecture Notes in Math. 370, Springer-Verlag 1973.

[23] Milne, J.S.,
The Tate-Šafarevič group of a constant abelian variety,
Invent. Math. 5 (1968), pp. 63-84.

[24] Milne, J.S.,
On a conjecture of Artin-Tate,
Ann. of Math. 102 (1975), pp. 517-553.

[25] Milne, J.S.,
Duality in the flat cohomology of a surface,
Ann. scient. Éc. Norm. Sup. 9 (1976), pp. 171-202.

[26] Milne, J.S.,
Values of zeta functions of varieties over finite fields,
Amer. J. Math. 108 (1986), pp. 267-360.

[27] Nygaard, N.,
Closedness of regular 1-forms on algebraic surfaces,
Ann. scient. Éc. Norm. Sup. 3e serié, t.12 (1979), pp. 33-45.

[28] Nygaard, N.,
Slopes of powers of Frobenius on crystalline cohomology,
Ann. scient. Éc. Norm. Sup. 4e série 14 (1981), pp. 369-401.

[29] Nygaard, N.,
The Tate conjecture for an ordinary K3 surface over finite fields,
Invent. Math. 74 (1983), pp. 213-237.

[30] Nygaard, N. and Ogus, A.,
The Tate conjecture for K3 surfaces of finite height,
Ann. of Math. 122 (1985), pp. 461-507.

[31] Oort, F.,
Subvarieties of moduli spaces,
Invent. Math. 24 (1974), pp. 95-119.

[32] Rudakov, A. H., and Shafarevich, I. R.,
Inseparable morphisms of algebraic surfaces,
Akad. Sc. SSSR. 40 (1976), pp. 1264-1307.

[33] Shioda, T.,
An example of unirational surfaces in characteristic p,
Math. Ann. 211 (1974), pp. 233-236.

[34] Shioda, T.,
Supersingular K3 surfaces,
Algebraic Geometry, Proc. Summer School Copenhagen, 1978, Lecture Notes in Math. 732, Springer-Verlag (1979), pp. 665-591.

[35] Shioda, T.,
On the Picard number of a Fermat surface,
J. Fac. Sci. Univ. Tokyo 28 (1982), pp. 175-184.

[36] Shioda, T.,
An explicit algorithm for computing the Picard number of certain algebraic surfaces,
Amer. J. Math. 108 (1986), pp. 415-432.

[37] Shioda, T.,
Some observations on Jacobi sums,
Advanced Studies in Pure Math. 12, North Holland-Kinokuniya (1987), pp. 119-135.

[38] Shioda, T., and Katsura, T.,
On Fermat varieties,
Tôhoku J. Math. 31 (1979), pp. 97-115.

[39] Suwa, N.,
De Rham cohomology of algebraic surfaces with $q = -p_a$ in characteristic $p$,
Proc. of the Japan-France conference in Algebraic Geometry, Tokyo-Kyoto, Lecture Notes in Math. 1016, Springer-Verlag (1983), pp. 73-85.

[40] Suwa, N., and Yui, N.,
Arithmetic of Fermat varieties I: Fermat motives and p-adic cohomology groups, MSRI Berkely Preprint 1988.

[41] Tate, J.,
Algebraic cycles and poles of zeta functions,
Arithmetic Algebraic Geometry,
Harper & Row (1965), pp. 93-110.

[42] Tate, J.,
Endomorphisms of abelian varieties over finite fields,
Invent. Math. 2 (1966), pp. 133-144.

[43] Tate, J.,
On the conjecture of Birch and Swinnerton-Dyer and a geometric analog,
Dix exposés sur la cohomologie des shémas, North-Holland, Amsterdam (1968), pp. 189-214.

[44] Waterhouse, W.,
Abelian varieties over finite fields,
Ann. scient. Éc. Norm. Sup. série 4, t.2, (1969), pp. 521-560.

[45] Weil, A.,
Numbers of solutions of equations in finite fields,
Bull. Amer. Math. Soc. 55 (1949), pp. 497-508.

[46] Weil, A.,
Jacobi sums as Grössencharaktere,
Trans. Amer. Math. Soc. 74 (1952), pp. 487-495.

[47] Yui, N.,
The arithmetic of products of two algebraic curves,
J. Alg. 98, No. 1 (1986), pp. 102-142;
Corrections, J. Alg. 109, No. 2 (1987), pp. 561.

Noriyuki SUWA
Department of Mathematics
Tokyo Denki University
Kanda-nishiki-cho 2-2
Chiyodaku, Tokyo, 101 JAPAN

Noriko YUI
Department of Mathematics and Statistics, Queen's University
Kingston, Ontario
CANADA K7L 3N6

Vol. 1232: P.C. Schuur, Asymptotic Analysis of Soliton Problems. VIII, 180 pages. 1986.

Vol. 1233: Stability Problems for Stochastic Models. Proceedings, 1985. Edited by V.V. Kalashnikov, B. Penkov and V.M. Zolotarev. VI, 223 pages. 1986.

Vol. 1234: Combinatoire énumérative. Proceedings, 1985. Edité par G. Labelle et P. Leroux. XIV, 387 pages. 1986.

Vol. 1235: Séminaire de Théorie du Potentiel, Paris, No. 8. Directeurs: M. Brelot, G. Choquet et J. Deny. Rédacteurs: F. Hirsch et G. Mokobodzki. III, 209 pages. 1987.

Vol. 1236: Stochastic Partial Differential Equations and Applications. Proceedings, 1985. Edited by G. Da Prato and L. Tubaro. V, 257 pages. 1987.

Vol. 1237: Rational Approximation and its Applications in Mathematics and Physics. Proceedings, 1985. Edited by J. Gilewicz, M. Pindor and W. Siemaszko. XII, 350 pages. 1987.

Vol. 1238: M. Holz, K.-P. Podewski and K. Steffens, Injective Choice Functions. VI, 183 pages. 1987.

Vol. 1239: P. Vojta, Diophantine Approximations and Value Distribution Theory. X, 132 pages. 1987.

Vol. 1240: Number Theory, New York 1984–85. Seminar. Edited by D.V. Chudnovsky, G.V. Chudnovsky, H. Cohn and M.B. Nathanson. V, 324 pages. 1987.

Vol. 1241: L. Gårding, Singularities in Linear Wave Propagation. III, 125 pages. 1987.

Vol. 1242: Functional Analysis II, with Contributions by J. Hoffmann-Jørgensen et al. Edited by S. Kurepa, H. Kraljević and D. Butković. VII, 432 pages. 1987.

Vol. 1243: Non Commutative Harmonic Analysis and Lie Groups. Proceedings, 1985. Edited by J. Carmona, P. Delorme and M. Vergne. V, 309 pages. 1987.

Vol. 1244: W. Müller, Manifolds with Cusps of Rank One. XI, 158 pages. 1987.

Vol. 1245: S. Rallis, L-Functions and the Oscillator Representation. XVI, 239 pages. 1987.

Vol. 1246: Hodge Theory. Proceedings, 1985. Edited by E. Cattani, F. Guillén, A. Kaplan and F. Puerta. VII, 175 pages. 1987.

Vol. 1247: Séminaire de Probabilités XXI. Proceedings. Edité par J. Azéma, P.A. Meyer et M. Yor. IV, 579 pages. 1987.

Vol. 1248: Nonlinear Semigroups, Partial Differential Equations and Attractors. Proceedings, 1985. Edited by T.L. Gill and W.W. Zachary. IX, 185 pages. 1987.

Vol. 1249: I. van den Berg, Nonstandard Asymptotic Analysis. IX, 187 pages. 1987.

Vol. 1250: Stochastic Processes – Mathematics and Physics II. Proceedings 1985. Edited by S. Albeverio, Ph. Blanchard and L. Streit. VI, 359 pages. 1987.

Vol. 1251: Differential Geometric Methods in Mathematical Physics. Proceedings, 1985. Edited by P.L. García and A. Pérez-Rendón. VII, 300 pages. 1987.

Vol. 1252: T. Kaise, Représentations de Weil et $GL_2$ Algèbres de division et $GL_n$. VII, 203 pages. 1987.

Vol. 1253: J. Fischer, An Approach to the Selberg Trace Formula via the Selberg Zeta-Function. III, 184 pages. 1987.

Vol. 1254: S. Gelbart, I. Piatetski-Shapiro, S. Rallis. Explicit Constructions of Automorphic L-Functions. VI, 152 pages. 1987.

Vol. 1255: Differential Geometry and Differential Equations. Proceedings, 1985. Edited by C. Gu, M. Berger and R.L. Bryant. XII, 243 pages. 1987.

Vol. 1256: Pseudo-Differential Operators. Proceedings, 1986. Edited by H.O. Cordes, B. Gramsch and H. Widom. X, 479 pages. 1987.

Vol. 1257: X. Wang, On the C*-Algebras of Foliations in the Plane. V, 165 pages. 1987.

Vol. 1258: J. Weidmann, Spectral Theory of Ordinary Differential Operators. VI, 303 pages. 1987.

Vol. 1259: F. Cano Torres, Desingularization Strategies for Three-Dimensional Vector Fields. IX, 189 pages. 1987.

Vol. 1260: N.H. Pavel, Nonlinear Evolution Operators and Semigroups. VI, 285 pages. 1987.

Vol. 1261: H. Abels, Finite Presentability of S-Arithmetic Groups. Compact Presentability of Solvable Groups. VI, 178 pages. 1987.

Vol. 1262: E. Hlawka (Hrsg.), Zahlentheoretische Analysis II. Seminar, 1984–86. V, 158 Seiten. 1987.

Vol. 1263: V.L. Hansen (Ed.), Differential Geometry. Proceedings, 1985. XI, 288 pages. 1987.

Vol. 1264: Wu Wen-tsün, Rational Homotopy Type. VIII, 219 pages. 1987.

Vol. 1265: W. Van Assche, Asymptotics for Orthogonal Polynomials. VI, 201 pages. 1987.

Vol. 1266: F. Ghione, C. Peskine, E. Sernesi (Eds.), Space Curves. Proceedings, 1985. VI, 272 pages. 1987.

Vol. 1267: J. Lindenstrauss, V.D. Milman (Eds.), Geometrical Aspects of Functional Analysis. Seminar. VII, 212 pages. 1987.

Vol. 1268: S.G. Krantz (Ed.), Complex Analysis. Seminar, 1986. VII, 195 pages. 1987.

Vol. 1269: M. Shiota, Nash Manifolds. VI, 223 pages. 1987.

Vol. 1270: C. Carasso, P.-A. Raviart, D. Serre (Eds.), Nonlinear Hyperbolic Problems. Proceedings, 1986. XV, 341 pages. 1987.

Vol. 1271: A.M. Cohen, W.H. Hesselink, W.L.J. van der Kallen, J.R. Strooker (Eds.), Algebraic Groups Utrecht 1986. Proceedings. XII, 284 pages. 1987.

Vol. 1272: M.S. Livšic, L.L. Waksman, Commuting Nonselfadjoint Operators in Hilbert Space. III, 115 pages. 1987.

Vol. 1273: G.-M. Greuel, G. Trautmann (Eds.), Singularities, Representation of Algebras, and Vector Bundles. Proceedings, 1985. XIV, 383 pages. 1987.

Vol. 1274: N. C. Phillips, Equivariant K-Theory and Freeness of Group Actions on C*-Algebras. VIII, 371 pages. 1987.

Vol. 1275: C.A. Berenstein (Ed.), Complex Analysis I. Proceedings, 1985–86. XV, 331 pages. 1987.

Vol. 1276: C.A. Berenstein (Ed.), Complex Analysis II. Proceedings, 1985–86. IX, 320 pages. 1987.

Vol. 1277: C.A. Berenstein (Ed.), Complex Analysis III. Proceedings, 1985–86. X, 350 pages. 1987.

Vol. 1278: S.S. Koh (Ed.), Invariant Theory. Proceedings, 1985. V, 102 pages. 1987.

Vol. 1279: D. Ieşan, Saint-Venant's Problem. VIII, 162 Seiten. 1987.

Vol. 1280: E. Neher, Jordan Triple Systems by the Grid Approach. XII, 193 pages. 1987.

Vol. 1281: O.H. Kegel, F. Menegazzo, G. Zacher (Eds.), Group Theory. Proceedings, 1986. VII, 179 pages. 1987.

Vol. 1282: D.E. Handelman, Positive Polynomials, Convex Integral Polytopes, and a Random Walk Problem. XI, 136 pages. 1987.

Vol. 1283: S. Mardešić, J. Segal (Eds.), Geometric Topology and Shape Theory. Proceedings, 1986. V, 261 pages. 1987.

Vol. 1284: B.H. Matzat, Konstruktive Galoistheorie. X, 286 pages. 1987.

Vol. 1285: I.W. Knowles, Y. Saitō (Eds.), Differential Equations and Mathematical Physics. Proceedings, 1986. XVI, 499 pages. 1987.

Vol. 1286: H.R. Miller, D.C. Ravenel (Eds.), Algebraic Topology. Proceedings, 1986. VII, 341 pages. 1987.

Vol. 1287: E.B. Saff (Ed.), Approximation Theory, Tampa. Proceedings, 1985–1986. V, 228 pages. 1987.

Vol. 1288: Yu. L. Rodin, Generalized Analytic Functions on Riemann Surfaces. V, 128 pages, 1987.

Vol. 1289: Yu. I. Manin (Ed.), K-Theory, Arithmetic and Geometry. Seminar, 1984–1986. V, 399 pages. 1987.

Vol. 1290: G. Wüstholz (Ed.), Diophantine Approximation and Transcendence Theory. Seminar, 1985. V, 243 pages. 1987.

Vol. 1291: C. Mœglin, M.-F. Vignéras, J.-L. Waldspurger, Correspondances de Howe sur un Corps p-adique. VII, 163 pages. 1987

Vol. 1292: J.T. Baldwin (Ed.), Classification Theory. Proceedings, 1985. VI, 500 pages. 1987.

Vol. 1293: W. Ebeling, The Monodromy Groups of Isolated Singularities of Complete Intersections. XIV, 153 pages. 1987.

Vol. 1294: M. Queffélec, Substitution Dynamical Systems – Spectral Analysis. XIII, 240 pages. 1987.

Vol. 1295: P. Lelong, P. Dolbeault, H. Skoda (Réd.), Séminaire d'Analyse P. Lelong – P. Dolbeault – H. Skoda. Seminar, 1985/1986. VII, 283 pages. 1987.

Vol. 1296: M.-P. Malliavin (Ed.), Séminaire d'Algèbre Paul Dubreil et Marie-Paule Malliavin. Proceedings, 1986. IV, 324 pages. 1987.

Vol. 1297: Zhu Y.-I., Guo B.-y. (Eds.), Numerical Methods for Partial Differential Equations. Proceedings. XI, 244 pages. 1987.

Vol. 1298: J. Aguadé, R. Kane (Eds.), Algebraic Topology, Barcelona 1986. Proceedings. X, 255 pages. 1987.

Vol. 1299: S. Watanabe, Yu.V. Prokhorov (Eds.), Probability Theory and Mathematical Statistics. Proceedings, 1986. VIII, 589 pages. 1988.

Vol. 1300: G.B. Seligman, Constructions of Lie Algebras and their Modules. VI, 190 pages. 1988.

Vol. 1301: N. Schappacher, Periods of Hecke Characters. XV, 160 pages. 1988.

Vol. 1302: M. Cwikel, J. Peetre, Y. Sagher, H. Wallin (Eds.), Function Spaces and Applications. Proceedings, 1986. VI, 445 pages. 1988.

Vol. 1303: L. Accardi, W. von Waldenfels (Eds.), Quantum Probability and Applications III. Proceedings, 1987. VI, 373 pages. 1988.

Vol. 1304: F.Q. Gouvêa, Arithmetic of *p*-adic Modular Forms. VIII, 121 pages. 1988.

Vol. 1305: D.S. Lubinsky, E.B. Saff, Strong Asymptotics for Extremal Polynomials Associated with Weights on ℝ. VII, 153 pages. 1988.

Vol. 1306: S.S. Chern (Ed.), Partial Differential Equations. Proceedings, 1986. VI, 294 pages. 1988.

Vol. 1307: T. Murai, A Real Variable Method for the Cauchy Transform, and Analytic Capacity. VIII, 133 pages. 1988.

Vol. 1308: P. Imkeller, Two-Parameter Martingales and Their Quadratic Variation. IV, 177 pages. 1988.

Vol. 1309: B. Fiedler, Global Bifurcation of Periodic Solutions with Symmetry. VIII, 144 pages. 1988.

Vol. 1310: O.A. Laudal, G. Pfister, Local Moduli and Singularities. V, 117 pages. 1988.

Vol. 1311: A. Holme, R. Speiser (Eds.), Algebraic Geometry, Sundance 1986. Proceedings. VI, 320 pages. 1988.

Vol. 1312: N.A. Shirokov, Analytic Functions Smooth up to the Boundary. III, 213 pages. 1988.

Vol. 1313: F. Colonius, Optimal Periodic Control. VI, 177 pages. 1988.

Vol. 1314: A. Futaki, Kähler-Einstein Metrics and Integral Invariants. IV, 140 pages. 1988.

Vol. 1315: R.A. McCoy, I. Ntantu, Topological Properties of Spaces of Continuous Functions. IV, 124 pages. 1988.

Vol. 1316: H. Korezlioglu, A.S. Ustunel (Eds.), Stochastic Analysis and Related Topics. Proceedings, 1986. V, 371 pages. 1988.

Vol. 1317: J. Lindenstrauss, V.D. Milman (Eds.), Geometric Aspects of Functional Analysis. Seminar, 1986–87. VII, 289 pages. 1988.

Vol. 1318: Y. Felix (Ed.), Algebraic Topology – Rational Homotopy. Proceedings, 1986. VIII, 245 pages. 1988

Vol. 1319: M. Vuorinen, Conformal Geometry and Quasiregular Mappings. XIX, 209 pages. 1988.

Vol. 1320: H. Jürgensen, G. Lallement, H.J. Weinert (Eds.), Semigroups, Theory and Applications. Proceedings, 1986. X, 416 pages. 1988.

Vol. 1321: J. Azéma, P.A. Meyer, M. Yor (Eds.), Séminaire de Probabilités XXII. Proceedings. IV, 600 pages. 1988.

Vol. 1322: M. Métivier, S. Watanabe (Eds.), Stochastic Analysis. Proceedings, 1987. VII, 197 pages. 1988.

Vol. 1323: D.R. Anderson, H.J. Munkholm, Boundedly Controlled Topology. XII, 309 pages. 1988.

Vol. 1324: F. Cardoso, D.G. de Figueiredo, R. Iório, O. Lopes (Eds.), Partial Differential Equations. Proceedings, 1986. VIII, 433 pages. 1988.

Vol. 1325: A. Truman, I.M. Davies (Eds.), Stochastic Mechanics and Stochastic Processes. Proceedings, 1986. V, 220 pages. 1988.

Vol. 1326: P.S. Landweber (Ed.), Elliptic Curves and Modular Forms in Algebraic Topology. Proceedings, 1986. V, 224 pages. 1988.

Vol. 1327: W. Bruns, U. Vetter, Determinantal Rings. VII,236 pages. 1988.

Vol. 1328: J.L. Bueso, P. Jara, B. Torrecillas (Eds.), Ring Theory. Proceedings, 1986. IX, 331 pages. 1988.

Vol. 1329: M. Alfaro, J.S. Dehesa, F.J. Marcellan, J.L. Rubio de Francia, J. Vinuesa (Eds.): Orthogonal Polynomials and their Applications. Proceedings, 1986. XV, 334 pages. 1988.

Vol. 1330: A. Ambrosetti, F. Gori, R. Lucchetti (Eds.), Mathematical Economics. Montecatini Terme 1986. Seminar. VII, 137 pages. 1988.

Vol. 1331: R. Bamón, R. Labarca, J. Palis Jr. (Eds.), Dynamical Systems, Valparaiso 1986. Proceedings. VI, 250 pages. 1988.

Vol. 1332: E. Odell, H. Rosenthal (Eds.), Functional Analysis. Proceedings, 1986–87. V, 202 pages. 1988.

Vol. 1333: A.S. Kechris, D.A. Martin, J.R. Steel (Eds.), Cabal Seminar 81–85. Proceedings, 1981–85. V, 224 pages. 1988.

Vol. 1334: Yu.G. Borisovich, Yu. E. Gliklikh (Eds.), Global Analysis – Studies and Applications III. V, 331 pages. 1988.

Vol. 1335: F. Guillén, V. Navarro Aznar, P. Pascual-Gainza, F. Puerta, Hyperrésolutions cubiques et descente cohomologique. XII, 192 pages. 1988.

Vol. 1336: B. Helffer, Semi-Classical Analysis for the Schrödinger Operator and Applications. V, 107 pages. 1988.

Vol. 1337: E. Sernesi (Ed.), Theory of Moduli. Seminar, 1985. VIII, 232 pages. 1988.

Vol. 1338: A.B. Mingarelli, S.G. Halvorsen, Non-Oscillation Domains of Differential Equations with Two Parameters. XI, 109 pages. 1988.

Vol. 1339: T. Sunada (Ed.), Geometry and Analysis of Manifolds. Procedings, 1987. IX, 277 pages. 1988.

Vol. 1340: S. Hildebrandt, D.S. Kinderlehrer, M. Miranda (Eds.), Calculus of Variations and Partial Differential Equations. Proceedings, 1986. IX, 301 pages. 1988.

Vol. 1341: M. Dauge, Elliptic Boundary Value Problems on Corner Domains. VIII, 259 pages. 1988.

Vol. 1342: J.C. Alexander (Ed.), Dynamical Systems. Proceedings, 1986–87. VIII, 726 pages. 1988.

Vol. 1343: H. Ulrich, Fixed Point Theory of Parametrized Equivariant Maps. VII, 147 pages. 1988.

Vol. 1344: J. Král, J. Lukeš, J. Netuka, J. Veselý (Eds.), Potential Theory – Surveys and Problems. Proceedings, 1987. VIII, 271 pages. 1988.

Vol. 1345: X. Gomez-Mont, J. Seade, A. Verjovski (Eds.), Holomorphic Dynamics. Proceedings, 1986. VII, 321 pages. 1988.

Vol. 1346: O. Ya. Viro (Ed.), Topology and Geometry – Rohlin Seminar. XI, 581 pages. 1988.

Vol. 1347: C. Preston, Iterates of Piecewise Monotone Mappings on an Interval. V, 166 pages. 1988.

Vol. 1348: F. Borceux (Ed.), Categorical Algebra and its Applications. Proceedings, 1987. VIII, 375 pages. 1988.

Vol. 1349: E. Novak, Deterministic and Stochastic Error Bounds in Numerical Analysis. V, 113 pages. 1988.